Lecture Notes in Electrical Engineering

Volume 252

For further volumes:
http://www.springer.com/series/7818

Xiao Liu · Qiang Xu

Trace-Based Post-Silicon Validation for VLSI Circuits

Springer

Xiao Liu
University of California, Berkeley
Berkeley, CA
USA

Qiang Xu
The Chinese University of Hong Kong
Shatin, N.T.
Hong Kong SAR

ISSN 1876-1100 ISSN 1876-1119 (electronic)
ISBN 978-3-319-37594-6 ISBN 978-3-319-00533-1 (eBook)
DOI 10.1007/978-3-319-00533-1
Springer Cham Heidelberg New York Dordrecht London

Printed on acid-free paper

Springer is part of Springer Science+Business Media (www.springer.com)

To my family

Preface

This book includes, but is not limited to, the research work on post-silicon validation during the author Xiao Liu's Ph.D. study. Post-silicon validation is an emerging research field, and limited publications are available to summarize the state of the art on it. Interested readers can refer to our survey paper [1] to know more background.

The book is roughly divided into three parts. The first part (Chaps. 1 and 2) covers the background of VLSI design trends, validation challenges, and current solutions that resolve various problems in post-silicon validation.

The major target of this book is how to conduct signal tracing effectively to provide sufficient observability and controllability during the debug process with low-cost design for debug (DfD) structures. Starting from Chap. 3, we present our automatic solutions to address the problem.

The second part (Chaps. 3–5) introduces several novel tracing solutions proposed in this book. The tracing techniques described in Chaps. 3 and 4 are for tackling functional error, by enhancing visibility and improving error detection quality, respectively, while Chap. 5 describes another tracing solution to deal with electrical error. In Chap. 3, based on the fact that logic values can be restored structurally, we define several evaluation metrics in a theoretically precise manner and conduct structural analysis to estimate visibility, which is then used to guide our trace signal selection algorithms for debugging functional errors. With this technique, the visibility can be dramatically enhanced by state restoration with selected trace data. The work in Chap. 3 is published in [8, 9]. Chapter 4 addresses the limitation that current tracing solutions constrain the debug capability by statically tracing the same signals in the circuit. Motivated by the above limitation, we propose a multiplexed signal tracing strategy. Within the strategy, we divide the tracing procedure into a few periods and develop the algorithms to selectively trace a different subset of accessible signals in each period, so that error detection quality will be improved. This work is published in [12]. Chapter 5 presents a tracing solution for tackling speedpath-related electrical error. This error only occurs under certain electrical environments and is extremely challenging to be root-caused. In our work, we first model the behavior of such electrical error. Accordingly, we propose a trace signal selection method for monitoring these errors. In addition, we develop a novel trace qualification technique that relies on reconfigurable logic to only store useful traced data, which improves the utilization of the trace buffer. To the best of our knowledge, this is the first tracing solution for debugging electrical errors in general logic circuits in post-silicon validation. With the technique, we are able to detect speedpath-related electrical errors at its root-caused site, on the exact error occurrence cycle, and without requiring any supporting "golden vector". The tracing solutions included in Chap. 5 are published in [11]. It is recommended to read Chap. 3 before Chaps. 4 and 5 because the probability-based terminologies defined in Chap. 3 are further extended to apply on the methods in Chaps. 4 and 5.

The third part (Chaps. 6–8) includes the solutions on cost-effective trace interconnection fabric design. The fabric introduced in Chap. 6 is for increasing the bandwidth of trace data by reusing existing test data transfer channel. The solution is further extended to avoid data corruption during multi-core debug [4]. Chapters 7 and 8 include different trace interconnection fabrics to facilitate flexible tracing and systematic tracing during post-silicon validation. To provide sufficient flexibility, we present a novel transfer structure and its optimization technique in Chap. 7. By combining multiplexor network and non-blocking network effectively, the technique can achieve high debug flexibility with minimized hardware cost. This work is published in [6]. On the other hand, we develop a low-cost fabric to support systematical error localization. As to be detailed in Chap. 8, with the fabric we can conduct an automatic tracing procedure to localize error with a few debug runs, so that the efficiency of error localization is greatly improved. Moreover, the fabric is inherently with high capability to tolerate unknown values in "golden vector". This work is published in [13]. The solutions proposed by Chaps. 6–8 are complementary ones so that the readers can freely read any chapter without going through others.

Every chapter in this book is largely self-contained. One thing to be noted is that the terminologies defined in each chapter are applicable for that chapter only.

Apart from the above automatic solutions on post-silicon validation included in this book, the author also conducted a series of works, which mainly focus on power relevant issues in manufacturing test, during his Ph.D. study. The author observes that growing test data volume and excessive test power are two of the major concerns for the industry when testing large integrated circuits. Various test data compression (TDC) schemes and low-power X-filling techniques were proposed to address the above problems. These methods, however, exploit the very same "don't-care" bits in the test cubes to achieve different objectives and hence may contradict each other. To address this problem, we propose several holistic solutions that target both issues together. In [2, 3, 5], we consider test power reduction in the code-based test compression environment. By studying and utilizing the features of different encoding techniques, our entropy-based generic solution is able to dramatically reduce test power with little compression ratio loss. For the linear decompressor-based TDC scheme widely used in the industry, we present a novel solution in [7] that dynamically extracts the relationship between don't-care bits in test patterns and then utilizes it to conduct effective X-filling for various power reduction objectives. These solutions outperform existing solutions significantly in test power reduction. Besides, we find that when testing delay faults on critical paths, conventional structural test patterns may be applied in functionally unreachable states, leading to over-testing or under testing of the circuits. In [10], we propose novel layout-aware pseudo-functional testing techniques to tackle the above problem. First, by taking the circuit layout information into account, functional constraints related to delay faults on critical paths are extracted. Then, we generate functionally reachable test cubes for every true critical path in the circuit. Finally, we fill the don't-care bits in the test cubes to maximize power supply noises on critical paths under the consideration of functional constraints. The technique is able to exercise the worst-case timing of critical paths in functional mode, which facilitates to reduce test escapes and test overkills simultaneously and guarantees high test quality.

To conclude, the solutions proposed during the author Xiao Liu's Ph.D. study aim at improving both validation and test quality on high reliable electronic devices.

Stanford, March 2013 Xiao Liu

References

Xu, Q. and Liu, X. (2010). On signal tracing in post-silicon validation. *In ProceedingsIEEE Asia South Pacific Design Automation Conference (ASP-DAC)*, (pp.262–267).

Li, J. Liu, X. Zhang, Y. Hu, Y. Li, X. and Xu, Q. (2008). On capture power-aware testdata compression for scan-based testing. *In Proceedings International Conferenceon Computer-Aided Design (ICCAD)*, (pp.67–72).

Li, J. Liu, X. Zhang, Y. Hu, Y. Li, X. and Xu, Q. (2011). Capture-power-aware testdata compression using selective encoding. *Integration, the VLSI journal, 44*(3).

Liu, X. and Xu, Q. (2008). On reusing test access mechanisms for debug data transferin SoC post-silicon validation. I*n Proceedings IEEE Asian Test Symposium (ATS*),(pp.303–308).

Liu, X. and Xu, Q. (2009). A generic framework for scan capture power reduction infixed-length symbol-based test compression environment. *In Proceedings Design,Automation, and Test in Europe (DATE*), (pp.1494–1499).

Liu, X. and Xu, Q. (2009). Interconnection fabric design for tracing signals in postsiliconvalidation. *In Proceedings ACM/IEEE Design Automation Conference(DAC*), (pp.352–357).

Liu, X. and Xu, Q. (2009). On simultaneous shift-and capture-power reduction in lineardecompressor-based test compression environment. *In Proceedings IEEE InternationalTest Conference (ITC*).

Liu, X. and Xu, Q. (2009). Trace signal selection for visibility enhancement in postsiliconvalidation. *In Proceedings Design, Automation, and Test in Europe (DATE*), (pp.1338–1343).

Liu, X. and Xu, Q. (2012).On signal selection for visibility enhancement in tracexprefacebased post-silicon validation. *IEEE Transactions on Computer-Aided Design ofIntegrated Circuits and Systems, 31* (pp.1263–1274).

Liu, X. Zhang, Y. Yuan, F. and Xu, Q. (2010). Layout-aware pseudo-functional testingfor critical paths considering power supply noise effects. *In Proceedings Design, Automation, and Test in Europe (DATE*), (pp.1432–1437).

Liu, X. and Xu, Q. (2010). On signal tracing for debugging speedpath-related electricalerrors in post-silicon validation. *In Proceedings IEEE Asian Test Symposium(ATS*), (pp.243–248).

Liu, X. and Xu, Q. (2011). On multiplexed signal tracing for post-silicon debug. *InProceedings Design, Automation, and Test in Europe (DATE*).

Liu, X. and Xu, Q. (2012). On efficient silicon debug with flexible and X-Toleranttrace interconnection fabric. *In Proceedings IEEE International Test Conference(ITC*).

Acknowledgments

I would like to thank my supervisor Professor Xu Qiang. In the past few years, he continually offered valuable guidance and taught me how to conduct solid research, how to write good research papers, and how to deliver clear presentations. All were essential to the completion of this thesis. Without doubt, the thesis would not have been possibly completed without his great effort.

I gratefully thank my markers, Professor Evangeline F. Y. Young, and Professor David Wu Yu Liang, whose constructive comments in my term presentations gave me important suggestions for my research progress.

I would also like to thank Professor Cheng for serving as external marker and reading my thesis.

I always feel lucky to work with my collaborators. To Yuan Feng and Zhang Yubin, I would like to deeply thank them for constantly sharing research ideas with me for the whole four years. To Tang Matthew, thank you for teaching me how to use the online resources of CSE department when I first came to the lab. I also acknowledge Ye Rong, Yu Haile, Zhang Jie, and Liu Yuxi for their advice and their willingness to share their thoughts with me. Thanks to my friends in the lab: Xiao Linfu, Qin Jing, Ma Qiang, Yang Xiaoqing, Jiang Yan, Jiang Mingqi, Qian Zaichen, Li Liang, and Tam Tak Kei, who make my life rich and colorful.

I was very fortunate in having so many helpful professors at Harbin Institute of Technology. I could never start my Ph.D. study without their prior guidance and unselfish support.

Without the support from my family, I would not be able to study in CUHK wholeheartedly. I would like to thank my mother for teaching me everything about engineering from my childhood. I also want to thank my father who strongly encouraged me when I was depressed.

Last but certainly not the least, I would like to thank my wife, Huang Lin. Without her unconditional support and encouragement, I could not possibly pursue the Ph.D. degree in CUHK. I really cherish the time we spent in Hong Kong together, and I cannot imagine how dull and difficult my life would be without her divine love and wisdom.

Contents

Chapter 1
Introduction

Due to the high design complexity and the inaccurate abstracted models used in today's VLSI design flow, pre-silicon verification process is not sufficient to eliminate all the errors. To tackle the remaining errors in the prototype silicon, in this book we introduce several innovative techniques to efficiently assist designers during post-silicon validation. Before delving into our solutions, we first review the trends of VLSI design and introduce various errors that occur in the design flow. We then summarize our validation techniques that resolve them.

1.1 VLSI Design Trends and Validation Challenges

The design flow of VLSI circuits includes three phases: specification, implementation and manufacturing. Designers describe the expected functionalities of the design in specification phase. They can use high-level abstraction languages like SystemC [22] to reduce the design complexity. At high level, the circuit is described as operations of data computations and transfers without the actual implementation of the components and interconnects. Designers can also use hardware description languages such as VHDL [50] or Verilog [51] to describe the design with more details. At the register-transfer level, the circuit is described by data flow and logical operations between registers. In implementation phase, computer-aided design tools are used to automatically transfer the above specification description into gate-level netlist (referred as logic synthesis), and then into physical layout with a few processing steps (e.g., floorplan, placement, routing, etc.). With the physical layout information, the circuit is fabricated in manufacturing phase. The fabrication process gradually generates the circuit with a series of photographic and chemical processing steps on semi-conducting material and mixture of other metals [19].

Modern VLSI circuits is designed to integrate more functionalities in each product. To facilitate this, circuits become more complex generation by generation. The trend can be demonstrated by the example of Intel's processors. In 1985, the 80386 processor only contains thousands of transistors, while the Core i5 processor released

X. Liu and Q. Xu, *Trace-Based Post-Silicon Validation for VLSI Circuits*, Lecture Notes in Electrical Engineering, DOI: 10.1007/978-3-319-00533-1_1,

in 2011 integrates millions of transistors; it is thousand times more powerful than the 80386 processor. The other trend in VLSI design flow is transistor becomes smaller by the development of integrated circuit (IC) manufacturing technology. The benefit is the transistors are able to switch faster and cost less power. On the other hand, with the technology scaling, it becomes more difficult, if not impossible, to obtain accurate model that represents the actual electrical behavior of the circuits.

Due to exponential design complexity and inaccurate abstracted models, it becomes extremely challenging to guarantee the correctness of the design. As a result, various errors exist to be tackled in the VLSI design flow. These errors can be classified as two types: functional error and electrical error. Functional error is from the mismatch between the design functional behavior and defined specification. Meanwhile, electrical error arises from the mismatches between circuit models in different abstraction levels, and the effects from new manufacturing technologies on signal integrity and process variation.

To ensure the functional correctness of the design, pre-silicon verification techniques including simulation and formal verification are extensively used in the implementation phase of VLSI design flow. Event-driven simulator is mostly used in simulation technique. It simulates the circuit behavior by dedicated and/or constrained random test vectors to verify the correctness of design's functionalities. The major drawback of simulation techniques is all possible behaviors of the design should be considered to conclude the design as 100 % error free. Consequently, as the number of possible behaviors increases exponentially with the circuit size, exhaustive simulation is impractical for modern circuits [58], and designers can only rely on the simulation techniques to achieve an acceptable coverage. Formal verification techniques get more attention in pre-silicon verification, as they are able to completely verify the correctness of design behaviors by using mathematical proofs [33, 37]. However, the computational complexity of these techniques is too high to apply on large circuits. In addition, since the design is verified against the specification in these techniques, the verification will not be correct when the reference specification contains some faults itself. By simulation and formal simulation, designers are able to detect and eliminate most design errors by pre-silicon verification. On the other hand, due to the high design complexity and ever-shrinking time-to-market window, it is difficult to guarantee the first silicon to be error free. Moreover, the strength of pre-silicon verification is further limited because it is not possible to model all physical characteristics of the design at high level abstraction [27].

Manufacturing test is an important technique for verifying the correctness of manufacturing phase in VLSI design flow [18]. It detects the mismatch between the fabricated design and the implemented one by testing the internal signal of the design with input patterns. Manufacturing test relies on the implemented design as the reference for detecting physical defects in the fabricated design. As a result, when some errors escape from pre-silicon verification and exist in the implemented design, they cannot be identified by the techniques of manufacturing test.

Physical probing techniques (such as, [49, 52, 60, 66]) are used to probe the internal signals and monitor their electrical behavior in the silicon. These techniques require expensive equipments and they are powerful for tackling electrical error.

Meanwhile, as physical probing techniques can only be applied on a very small suspicious region of the circuit, designers are required to conduct effective error localization method (e.g., the solution introduced in Chap. 5) before applying physical probing techniques.

As above techniques cannot guarantee to completely eliminate all the design errors before tape-out, today's complex circuits usually need to go through one or more re-spins to become bug-free [24, 68], even though half of the system development effort is allocated to verification tasks [61]. Since time-to-market dictates the success of a chip, it is vital to efficiently resolve the escaped errors in first silicon [2]. This step is known as *post-silicon validation* in the design flow. The main difficulty in post-silicon validation lies in the limited visibility and controllability for circuit internal nodes, because the circuit is a piece of silicon that has already been fabricated. Hence, several solutions are proposed in this research area to address the problem. In Chap. 2 we will briefly summarize the state of the art of post-silicon validation.

1.2 Key Contributions and Book Outline

In this book we present advanced technologies that target on improving the observability and controllability in trace-based post-silicon validation so that different types of errors are resolved efficiently.

By surveying the current techniques on trace-based post-silicon validation, we observe that how to select the trace signals is the primary problem that determines the effectiveness of trace-based debug strategy. Because if we select the right signals to trace so that error leaves "evidences" on them, localizing the error will be easy. Otherwise, the debug process can be quite challenging. In current practice, designers select those signals manually based on their experience. This ad-hoc process cannot guarantee the quality in debug process. More importantly, errors often occur in unexpected scenarios and it is impossible to predict which signals will be related to them during the design phase. From this aspect, our first work (i.e., Chap. 3) proposes an automated trace signal selection strategy so that the visibility is dramatically enhanced when debugging functional design error. Superior than previous work [35], we introduce theoretically accurate probability-based metrics to guide the signal selection procedure. Moreover, to apply this technique in practice, we further extend the signal selection technique to work with "golden vectors" from high-level simulation and assertion-based debug.

Despite the trace signals are carefully selected by the technique in Chap. 3, current trace-based debug solutions have another limitation. That is, they typically trace the same set of selected signals in each debug run and this kind of "static" tracing methods limits the error detection capability of the design. By leveraging the fact that the number of tapped signals is much more than the number of signals that can be traced concurrently, we propose a multiplexed signal tracing strategy in Chap. 4. In this tracing strategy, we divide the whole tracing procedure in each debug run into a few periods and intelligently trace different sets of tapped signals in each period. Within the tracing strategy, we introduce the signal grouping algorithm and supporting hardware design, so that the tracing strategy will greatly improve error detection capability.

The tracing solutions introduced in Chaps. 3 and 4 are applicable for resolving functional design errors. We then propose a tracing solution for tackling electrical errors in Chap. 5. Electrical errors are extremely challenging to be repeated and root-caused as they only manifest in certain electrical environment. Physical probing techniques can effectively eliminate these errors by monitoring the electrical behavior of internal node in the silicon. Meanwhile, designers have to rely on other strategies to detect and localize the error in a small region before using these techniques. To facilitate it, we propose a tracing solution to monitor those internal signals that are related to its electrical abnormal behavior. To be specific, we first model the behavior of electrical error. Next we propose a signal selection method to maximize the error detection probability. In addition, we develop a novel trace qualification hardware to improve the utilization of trace buffer. With this technique, designers are able to detect electrical errors at its root-caused site, on the error occurrence cycle and without requiring any supporting "golden vectors".

In Chaps. 6–8, we focus on developing the interconnection fabrics that transfer trace data. To provide sufficient debug capability, these fabrics are required to provide high data volume and high flexibility during trace data transfer. On the other hand, they should be designed with low cost because they are totally useless to users. Chapter 6 addresses the problem of limited debug access bandwidth in the interconnection fabric. By observing the circuits often contain dedicated test access mechanisms to transfer test data, we propose to reuse these resources for real-time trace data transfer in post-silicon validation. This strategy significantly increases data bandwidth with negligible routing overhead. In addition, we design DfD structures applicable for multi-core debug.

The debug capability in post-silicon validation depends on the flexibility during trace data transfer. Existing solutions typically use pipelined multiplexer trees to conduct the transfer duty. As more than one signal through the multiplexer cannot be observed concurrently, this ad-hoc technique limits the flexibility in debug process. In Chap. 7, we develop a novel interconnection fabric design to tackle the problem. With the fabric, designers are able to flexibly trace any signal combinations in each debug run, so that the debug capability is dramatically enhanced. To keep the fabric at low DfD hardware cost, we introduce simplification techniques on both multiplexer network and non-blocking concentration network within the fabric.

High flexibility is obtained by using the fabric introduced by Chap. 7. However, the provided flexibility cannot directly improve debug capability with associated tracing solution. In Chap. 8, we propose a different trace interconnection fabric design with little DfD hardware to achieve this. With the fabric, we introduce a systematic signal tracing procedure to automatically localize erroneous signals with just a few debug runs, so that the debug efficiency is greatly improved. Besides, the fabric is demonstrated to tolerate unknown bits in "golden vectors" during the tracing procedure.

Chapter 9 finally summarizes the contribution of this book and points out the directions for future work.

Chapter 2
State of the Art on Post-Silicon Validation

The first widely-adopted post-silicon validation technique utilized by the industry is to reuse the IEEE Std. 1149.1 (JTAG) test access port and existing design for test (DfT) structures in the circuit (e.g., scan chains) to run, halt and step the circuit under debug (CUD) to find bugs [71]. This technique is quite effective in identifying those easy-to-find bugs that leave "evidences" when the circuit halts, but fails to find those tricky bugs that manifest themselves only after a long period of operational time [65]. In addition, the behavior of many bugs is hard to repeat, making diagnosis with this run/stop debug methodology even more difficult. To mitigate the above problem, designers can add shadow flip-flops (FFs) to the CUD to increase its visibility during normal operation [32]. However, this method can only sample a few snapshots of the circuit's operational states and it also involves nontrivial DfD overhead.

To be able to root-cause design bugs, post-silicon validation requires to increase controllability and observability of the CUD's internal behavior to a much higher level than what manufacturing test generally needs [24]. A more effective silicon debug technique is to selectively monitor and trace internal signals of the circuit continuously during its normal operation [3]. The traced data can then be either stored in an on-chip trace buffer or transferred out of the chip via a trace port for later analysis.

The hardware infrastructure to facilitate trace-based silicon debug is shown in Fig. 2.1, wherein various DfD modules are introduced at design stage of the CUD for later debug purpose. Generally speaking, designers select to tap a number of signals in the CUD (typically thousands of signals in million-gate industrial designs [1]). However, only a subset of the tapped signals are traced concurrently during debug phase due to trace bandwidth limitation. This is achieved by an "interconnection fabric" (e.g., a MUX tree) that links the tapped signals to trace buffers or trace ports. In addition, trigger units are typically used to determine when to start and stop signal tracing so as to further reduce trace bandwidth requirement. In most cases, designers reuse JTAG test access port as the control interface for the debug phase.

When the first silicon is back with some bugs, in each debug run (see Fig. 2.2), designers first configure the DfD module in the CUD by selecting the to-be-traced

X. Liu and Q. Xu, *Trace-Based Post-Silicon Validation for VLSI Circuits*,
Lecture Notes in Electrical Engineering, DOI: 10.1007/978-3-319-00533-1_2,

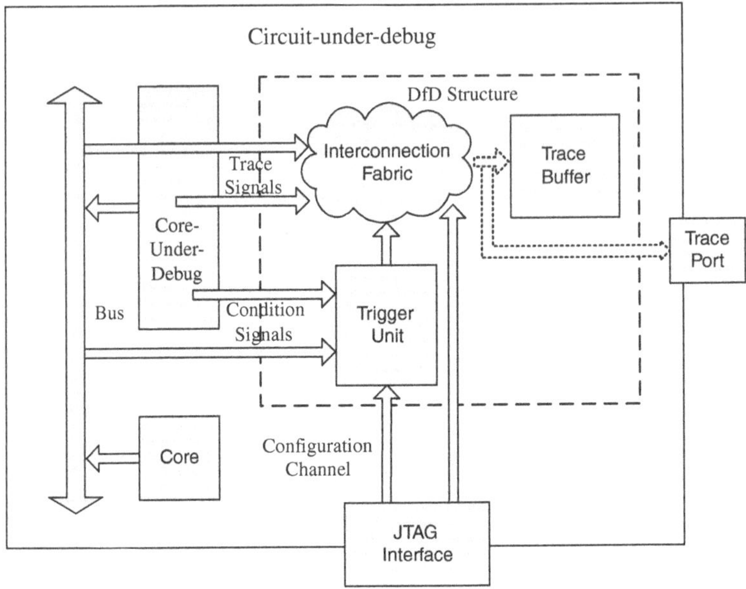

Fig. 2.1 Trace-based debug infrastructure [75]

signals from the tapped ones and determining the trigger conditions for signal tracing, and then put the CUD into normal operational mode. If the pre-determined trace condition is met, the traced data is transferred through the interconnection fabric to on-chip trace buffers or off-chip trace ports. The collected data are then analyzed to root-cause the possible design bugs. The above process iterates a number of debug rounds or even a few re-spins (when unlucky), until all the bugs are eliminated.

Figure 2.3 depicts the ARM *CoreSight* trace-based debug solution [7], wherein each ARM core is equipped with an embedded trace macrocell (ETM) for capturing the processor's states. It also contains a cross trigger interface so that trigger events can be transferred between different ETMs to facilitate multi-core debug.

A huge volume of trace data, however, is difficult to analyze and results in high DfD overhead. Therefore, how to conduct signal tracing effectively for bug elimination while keeping the associated hardware cost manageable (usually required to be less than 10 % of the original design) is a challenging task for IC designers. In the following we provide in-depth discussion for trace-based debug strategy and review recent advancements in this important area. In particular, we discuss the following issues in trace-based debug solution:

- Out of the large amount of state elements in the circuit, which signals should we choose to monitor and trace to provide high visibility to the CUD?
- How do we design the trace data transfer module to provide enough observation flexibility while keeping the associated DfD overhead low?

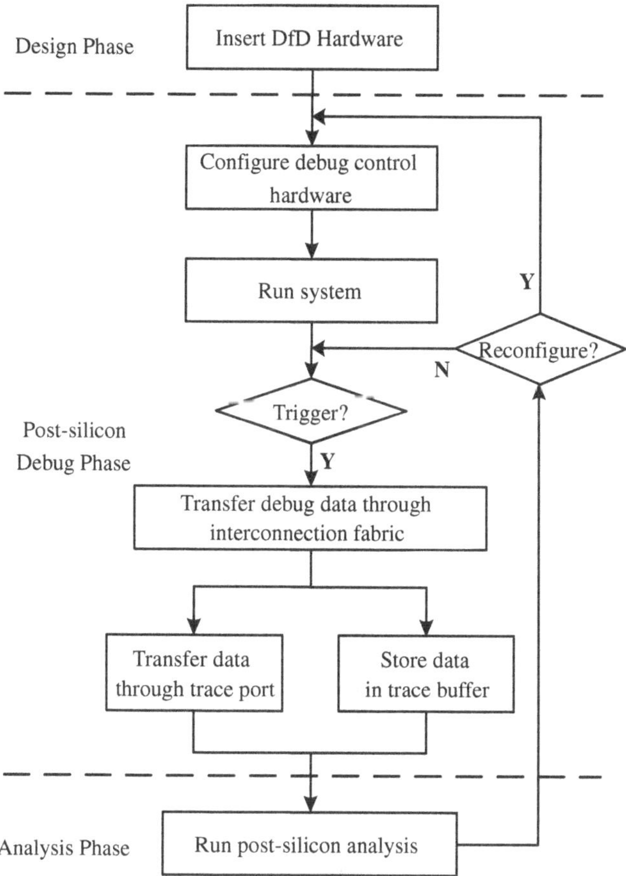

Fig. 2.2 Trace-based silicon debug flow

- How can we compress the large volume of trace data effectively so as to make efficient use of the limited trace bandwidth provided by trace buffers and/or trace ports?
- How do we control the signal tracing effectively to obtain highly-qualified trace data?

2.1 Trace Signal Selection

Trace signal selection becomes the primary problem in trace-based debug. This is because, ideally we wish to "see any signal at any time" during post-silicon validation and clearly it is not achievable with the large amount of internal signals deeply embedded in the fabricated chip. As indicated in Fig. 2.1, with the help of constrained

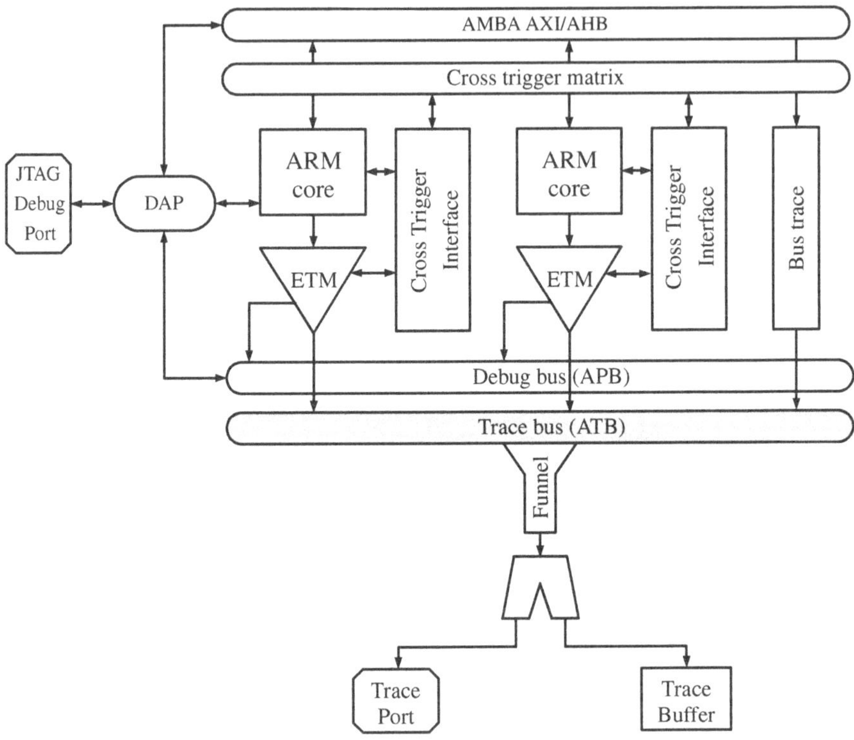

Fig. 2.3 ARM coreSight multi-core debug architecture [7]

DfD resources, we can only afford to tap a few internal state elements and use them to help designers root-cause the abnormal behaviors of the CUD. The objective is therefore to select those *essential* signals in the CUD so that bugs have a high chance to leave "evidences" on them, and the effectiveness of trace-based debug strategy highly relies on which signals are selected to be traced.

To debug errors on microprocessors and software running on them, naturally it is beneficial to observe the execution of the instructions. In [53], the authors proposed to trace the behavior of every execution stage of instructions to obtain more detailed information on how the microprocessor operates. In addition, several methods have been presented to monitor either the communication interface of the processor (data channel, address channel and control channel) [7, 39, 64], or the memory contents that store the execution results [38, 73].

For the increasingly popular network-on-chip (NoC) based designs, more visibilities are required to the communication among multiple cores, especially at transaction level to provide a globally consistent view of the system. Several trace selection methods were proposed for such type of designs in the literature to tackle this problem [17, 65, 72]. As an example, [17] proposed to attach dedicated monitoring probes on routers and provide transaction level observability of the NoC.

The above techniques are quite effective for the targeted types of circuits, but we are still facing the trace signal selection problem for general logic circuitries. In current practice, designers usually manually select those signals that are considered to be vulnerable to bugs or important for analysis to trace, based on their own design experience. This ad-hoc method, however, cannot guarantee the quality of the selected trace signals. More importantly, bugs often occur in unexpected scenarios and it is very difficult, if not impossible, to predict which signals will be related to them during the design phase. Therefore, we need to have at least some trace signals that are selected in an automated manner without designers' intervention.

2.2 Interconnection Fabric Design for Trace Data Transfer

As designers are not knowledgeable about which part of the design may contain bugs, a relatively large number of signals are selected to be traceable in the circuit, typically in the thousand range for million-gate designs [1]. Due to the associated DfD area cost and debug bandwidth requirement, however, it is impossible to *concurrently* monitor and trace all the tapped signals. Instead, only a small number of internal signals can be real-time observed together, and it is up to the designers to determine which signals to trace at a specific debug run, according to the system's erroneous behavior. These signals are then transferred to on-chip trace buffers and/or off-chip trace ports for diagnosis.

To reduce the DfD cost, industrial designs typically use MUX trees to select a subset of the tapped signals to trace in each debug run, in which the control signals to the multiplexers are configured through the JTAG interface (e.g., [1, 69]). To meet timing constraint for the tracing logic, the MUX trees can be pipelined. In addition, when the tapped signals come from multiple clock domains, first-in first-out (FIFO) buffers and/or flip-flop chains can be used to ensure data safety [1]. The above design methodology, however, limits this flexibility of observing any combinations of related tapped signals and reduces the visibility to the CUD, as any signals going through the same multiplexer cannot be traced concurrently. This problem can be easily solved by introducing non-blocking concentration network [48], which is able to select any m signals out of n inputs ($m \leq n$) and output them to the trace buffers/ports, but such design is with prohibitive DfD cost.

2.3 Trace Data Compression

Due to the limited trace bandwidth provided by trace buffers and/or trace ports, storing the "raw" traced data is not quite economical. Various trace data compression methodologies were presented to tackle this problem effectively.

In [53], the authors utilized the locality feature of instruction sequence and redundant information in monitored data that can be easily identified with the executed instruction to store the execution states of microprocessor. With the technique, a small amount of *footprints* are enough to observe the whole operational behavior of the microprocessor under debug. Recently, several works [38, 73] were presented for tracing the contents in cache. They both utilize the data locality feature when accessing cache and adopt dictionary-based compression to further improve the compression ratio. Different from each other, [73] observed the following features to enhance compression ratio: the similarity in tag field caused by spacial locality of memory reference and the unusual usage of high order bits for integer value; [38], on the other hand, proposed to reuse cache to compress instructions by inserting supporting module into it. The method is able to restore full information with a small amount of traced data combined with the contents remaining in cache.

Recently, a lossy compression method based on multiple-input signature register (MISR) was presented in [4]. With the assumption that the CUD behaves repeatable in different debug iterations, the method consecutively zooms-in the sampling intervals with compressed failure signatures generated from MISR to localize the error.

2.4 Trace-Based Debug Control

As shown in Fig. 2.1, trigger unit is designed for determining the signal tracing behavior. The behavior can be start and stop tracing so that unnecessary data can be filtered to reduce trace data volume (known as *trace qualification*). More importantly, the designers' capability of controlling the CUD directly affects the debug effectiveness. This is because, when designers can easily control the CUD into suspicious state and obtain relevant information, it becomes much earlier for them to root-cause the possible errors. The problem has been extensively studied in academia.

In [71], the authors described several basic DfD modules that can be used for debug control, including comparator to check if the condition signals are met with pre-configured value, and counter to facilitate the trigger control with temporal information. Later, [10] proposed to synthesize more complex control unit (i.e., assertion checker) to monitor complex behaviors of the CUD (e.g., ATB communication protocol). The unit is a state machine that can be generated from the description with formal languages for assertion-based verification. Later, the same authors [11] introduced several enhanced features to localize the errors that are buried as internal states in sophisticated assertions, in addition to reduce the associated hardware cost in unit generation in [10].

Multi-core debug control for complex SoC devices is a challenging task since we need to be able to control related cores simultaneously [72]. This problem becomes particularly difficult when the data transfer among cores is not deterministic, in which case, it is rather ineffective to configure all the required trigger conditions before running the system. To tackle this problem, [65] proposed a so-called in-band cross-trigger event transmission infrastructure. By inserting the cross-trigger events into the messages, designers are able to trace the desired messages more easily.

Chapter 3
Signal Selection for Visibility Enhancement

As stated in Sect. 2.1, the effectiveness of trace-based debug strategy highly relies on which signals are selected to trace in the system. This is because, if we select the right signals to trace so that a bug leaves "evidences" on them, finding the root cause for the bug would be quite easy. Otherwise, the debug process can be quite challenging. In current practice, designers *manually* select those signals that are considered to be vulnerable to bugs or important for analysis to trace, based on their design experience. While their knowledge about the design is of great help in trace signal selection, this ad-hoc process cannot guarantee the quality of the selected trace signals. More importantly, bugs often occur in unexpected scenarios and it is impossible to predict which signals will be related to them during the design phase. From this aspect, to achieve effective bug detection, we propose an automated trace signal selection strategy with a new probability-based evaluation metric, with which the visibility is dramatically enhanced when debugging functional design error.

In the remainder of this chapter, Sect. 3.1 reviews related prior work on trace-based post-silicon validation and then motivates this work. The proposed gate-level restorability definitions and our automated trace signal selection methodologies are illustrated in detail in Sects. 3.2 and 3.3, respectively. In Sect. 3.4, we present our experimental results on benchmark circuits. Finally, Sect. 3.5 concludes this work.

3.1 Preliminaries and Summary of Contributions

With the help of limited DfD resources, it is possible to trace a few signals in the system to view parts of the system state and use them to reason the root cause of the design's erroneous behavior. Since the number of signals that can be concurrently traced is constrained, the key question is then how to select them intelligently to identify challenging bugs effectively. For processors in a system, we can simply select those signals that are enough for us to reconstruct the program flow [53]. For random logic, however, this is a rather difficult problem and designers typically select

X. Liu and Q. Xu, *Trace-Based Post-Silicon Validation for VLSI Circuits*, Lecture Notes in Electrical Engineering, DOI: 10.1007/978-3-319-00533-1_3,

trace signals manually according to their own experience and their knowledge about the design.

As bugs often occur in unexpected scenarios, it is important to have some trace signals that are selected in an automated manner without designers' intervention. Ko and Nicolici [35] is the first attempt to address the above problem in the literature. In this work, the authors observed that it is possible to expand the logic states on the few trace signals to restore many missing states on those untraced signals by conducting structural analysis on the circuit.[1] With the restored states and traced states, designers can compare them with the "golden vector" (possibly from high-level simulation) to detect the erroneous states, based on which the logic diagnosis solutions (e.g., [15]) can be further utilized to localize the root-caused design error. Later, several other signal selection techniques were presented in the literature to maximize restoration capability (e.g., [9, 36, 42, 62]). Similar to [35], probability-based evaluation metrics were employed to model signals' restoration capability. Prabhakar and Hsiao [55] introduced an implication-based approach for trace signal selection, wherein signal correlation is obtained via recursive learning and used to improve solution quality. This technique, however, is quite time-consuming. In [78], the authors proposed a trace signal selection method, targeting on a different objective, minimizing the latency for capturing the propagated evidences after bugs are triggered. This work assumes that the states' sequence for exposing error is available, which may not be the case at early design stage.

In [35], The *restoration ratio*, calculated as $R = \frac{N_{\text{traced}}+N_{\text{restored}}}{N_{\text{traced}}}$, is used as the evaluation metric to measure the quality of the selected trace signals. N_{traced} and N_{restored} represent the number of traced states and that of restored states, respectively. The state restoration process is conducted as follows. Consider a 2-input AND gate. If one of the inputs is '0', the output c can be inferred as '0' through forward propagation. Meanwhile, when the output c is '1', the two inputs a and b can be derived as '1' by backward justification. Based on the above simple operations, for an example sequential circuit in Fig. 3.1a, when flip-flop (FF) C is traced for four clock cycles (Cycle 0 to Cycle 4), the missing states can be inferred as shown in Fig. 3.1b. The restoration ratio is hence $R = 14/4 = 3.5$ ($N_{\text{traced}} = 4$, $N_{\text{restored}} = 10$). It is important to note that the above state restoration is correct under the assumption that there are no timing errors between the traced signals and the restored signals (designers can vary the circuit operational speed during the debug process to make it true, if necessary), and hence is only effective for debugging functional errors.

Before observing the actual values on the traced signals from the prototype CUD, during the design phase, we can only estimate the effectiveness (referred as restorability) of each trace signal. Ko and Nicolici [35] defined the *forward restorability* and *backward restorability* for logic gates as shown in Fig. 3.2. Based on their definition, the calculated restoration ratio equals $10/4 = 2.5$.

The above definitions for the gate-level restorabilities, however, are lack of theoretical basis and hence are not accurate in many cases. Still consider a 2-input AND

[1] Similar concept has been mentioned earlier in [26] as "data expansion" technique, but for combinational logic only.

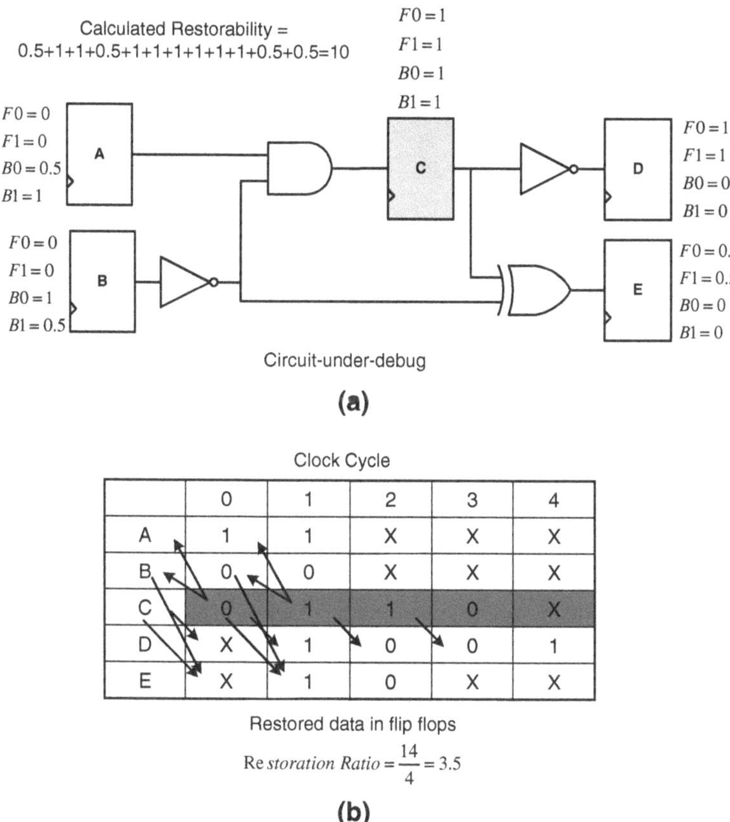

	0	1	2	3	4
A	1	1	X	X	X
B	0	0	X	X	X
C	0	1	1	0	X
D	X	1	0	0	1
E	X	1	0	X	X

Fig. 3.1 State restoration example in [35]

gate with two inputs a and b. Suppose we cannot restore the output $c = 0$ (i.e., $B0(c) = 0$), it is obvious that $a = 0$ cannot be restored through backward justification and $B0(a)$ should equal to 0. However, with the formula shown in Fig. 3.2, we have $B0(a) = (\max\{B0(c)\} + F1(b))/2 = F1(b)/2$. Since the value of $F1(b)$ is dependent on the restorabilities of its driving gates, it can be any value between 0 and 1. The restorability calculated in [35] is hence inaccurate and it can be easily propagated to a large number of circuit nodes, which makes the circuit-level visibility calculation quite time-consuming. In addition, during the restorability calculation in [35], forward propagation and backward justification are applied separately to find missing states, which may under-estimate the possible restored states. As shown in Fig. 3.1, the state value '0' of FF B in Cycle 1 is restored backward by the state value '1' of FF C in Cycle 2. Then the newly restored value, together with traced value '1' of FF C in Cycle 1, can forwardly restore the state value '0' of FF E in Cycle 2. Hence, these two directions of restorability calculation should be considered relatively.

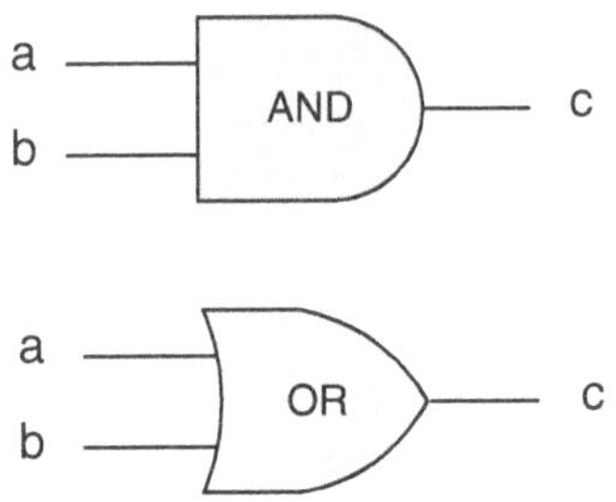

a
b
AND
c

$$F0(c) = \max\{F0(a), F0(b)\}$$
$$F1(c) = (F1(a) + F1(b))/2$$
$$B0(a) = (\max\{B0(c)\} + F1(b))/2$$
$$B1(a) = \max\{B1(c)\}$$

a
b
OR
c

$$F0(c) = (F0(a) + F0(b))/2$$
$$F1(c) = \max\{F1(a), F1(b)\}$$
$$B0(a) = \max\{B0(c)\}$$
$$B1(a) = (\max\{B1(c)\} + F0(b))/2$$

F1/F0 --- the probability of restoring data 1/0 of a node through forward propagation
B1/B0 --- the probability of restoring data 1/0 of a node through backward justification

Fig. 3.2 Forward/backward restorability definitions in [35]

The limitations in [35] motivate us to propose new automated trace signal selection methodologies for improving the visibility. In addition, to apply the methodology in the real applications of silicon debug, after selected data is traced and restored, we need to compare it with the "golden vector" for detecting erroneous state. However, as the "golden vector" is possibly obtained from high-level simulation, only part of state signals are known for part of cycles within the tracing process. Hence our solution should be extended to tackle the problem. On the other hand, in assertion-based debug, designers will pre-define the assertions that they are particularly interested in and we should also enhance our methodology to address the relevant issue. In this chapter, we present novel solutions to address the above problems. The main contributions include

- we define the gate-level restorabilities for visibility enhancement in a theoretically-precise manner;
- we propose novel automated trace signal selection algorithms that are able to restore much more missing states when compared to [35];
- we enhance the trace signal selection algorithm with new introduced metrics to work with high-level simulation and assertion-based debug.

3.2 Restorability Formulation

To calculate the restorability precisely, we first define the terminologies as shown in Table 3.1 and then we illustrate our probability-based calculations for gate-level restorabilities.

Table 3.1 Terminologies for restorability calculation

Selected visibility ($SV1/SV0$)	For selected trace nodes, $SV1 = SV0 = 1$; otherwise $SV1 = SV0 = 0$
Forward restorability ($FR1/FR0$)	The **conditional** probability to be restored as '1'/'0' by forward propagation **when** it is '1'/'0'
Backward restorability ($BR1/BR0$)	The **conditional** probability to be restored as '1'/'0' by backward justification **when** it is '1'/'0'
Restorability ($R1/R0$)	The **conditional** probability that the node is restored as '1'/'0' from other nodes **when** it is '1'/'0'
Functional probability ($P1/P0$)	The probability that the node is '1'/'0'in functional mode
Restored visibility ($RV1/RV0$)	The probability that the restored value '1'/'0' is **actually** observed on the node
Visibility ($V1/V0$)	The probability that the value '1'/'0' is **actually** observed on the node

3.2.1 Terminologies

It is obvious the visibility for those signals that are selected to trace is 1 otherwise it is 0, and we use $SV1/SV0$ to denote it. For the untraced signals, they can be restored by two *independent* methods: forward propagation and backward justification, and we use $FR1/FR0$ and $BR1/BR0$ to represent the probabilities to restore them by the corresponding method. Consequently, the total possibility that the value '1'/'0' can be restored on the untraced signals, denoted as $R1/R0$, can be calculated as:

$$R1 = FR1 + BR1 - FR1 \times BR1 \tag{3.1}$$

$$R0 = FR0 + BR0 - FR0 \times BR0 \tag{3.2}$$

It is important to note that the above restorabilities are conditional probabilities and they do not represent the probability that the value '1'/'0' is *actually* observed on the node. For example, if we are able to fully restore value '1' on a particular node but cannot restore '0' on it (i.e., $R1 = 1$ and $R0 = 0$), and suppose this node stuck at logic '0' all the time during the silicon debug procedure, we actually cannot observe anything for this node. To address this problem, we need to consider the probability for a node to be '1'/'0' in functional mode (represented by $P1/P0$) and we calculate the restored visibility for this node as:

$$RV1 = R1 \times P1 \tag{3.3}$$

$$RV0 = R0 \times P0 \tag{3.4}$$

Several methods can be utilized to obtain the signal probability for CUD in its functional mode. If the test sequence is pre-decided by designers, we can collect the statistical value by running the simulation and dumping internal states. Otherwise, we can simply calculate the probability with the method proposed in [12] by assuming some control inputs are pre-set as '1'/'0' to insure the CUD is working in functional mode, while all other inputs are with the probability 0.5 to be value '1'/'0'.

Finally, we use $V = \max\{SV, RV\}$ to unify the two kinds of visibilities obtained through direct trace or indirect restoration.

3.2.2 Gate-Level Restorabilities

Let us first take the two-input AND gate as an example to explain our definition for gate-level restorability first. For forward propagation, the output (c) is '1' only when both inputs (a and b) are '1'; while it is '0' when no less than one input is '0'. For backward justification, both inputs (a and b) are '1' when output (c) is '1', and one input (e.g., a) is '0' only when the output c is '0' and at the same time the other input (b) is '1'. Based on the above, we define the restorabilities as follows.

$$FR1(c) = \frac{V1(a) \bigcap V1(b)}{P1(c)} \tag{3.5}$$

$$FR0(c) = \frac{V0(a) \bigcup V0(b)}{P0(c)} \tag{3.6}$$

$$BR1(a) = \frac{V1(c)}{P1(a)} \tag{3.7}$$

$$BR0(a) = \frac{V0(c) \bigcap V1(b)}{P0(a)} \tag{3.8}$$

Due to the computational complexity, however, it is not possible to record the precise relationship between every signal. Therefore, we assume the inputs to every gate are *independent*. Equations (3.5) and (3.6) can then be simplified as follows.

$$FR1(c) = \frac{V1(a) \times V1(b)}{P1(a) \times P1(b)} \tag{3.9}$$

$$FR0(c) = \frac{V0(a) + V0(b) - V0(a) \times V0(b)}{1 - P1(a) \times P1(b)} \tag{3.10}$$

To calculate $BR0(a)$ within the corresponding event $V0(c) \bigcap V1(b)$, let us first consider the case that b and c are not traced. We cannot obtain $V0(c) \times V1(b) = RV0(c) \times V1(b) = P0(c) \times R0(c) \times V1(b)$ because the probability value on signal c strongly depends on the value on signal b. Fortunately, we notice that due to the nature of two-input AND gate, state $a = 0$ can be backward restored only in the event of *a*=*0, b*=*1, c*=*0*. The accurate probability for this event is $P0(a) \times P1(b)$ (see Table 3.2). Therefore, $V0(c) \bigcap V1(b)$ should be $\big(P0(a) \times P1(b)\big) \times \big(R0(c) \times R1(b)\big) = P0(a) \times R0(c) \times V1(b)$. Substituting it into Eq. (3.8) yields

Table 3.2 Example for AND gate backward justification

Event	Probability	c	b	a (truly)	a (restored)
$a = 1, b = 0, c = 0$	$P1(a) \times P0(b)$	0	0	1	X
$a = 0, b = 1, c = 0$	$P0(a) \times P1(b)$	0	1	0	0
$a = 0, b = 0, c = 0$	$P0(a) \times P0(b)$	0	0	0	X
$a = 1, b = 1, c = 1$	$P1(a) \times P1(b)$	1	1	1	1

$$BR0(a)|_{c \text{ is not traced}} = \frac{P0(a) \times R0(c) \times V1(b)}{P0(a)} = R0(c) \times V1(b) \quad (3.11)$$

For the case when c is a traced signal, it is observable all the time and we can restore $R0(a)$ as long as $b = 1$ is visible. Therefore,

$$BR0(a)|_{c \text{ is traced}} = V1(b) \quad (3.12)$$

It can be verified that Eqs. (3.11) and (3.12) also hold when b is traced. Based on the above principles, the restorability calculations for OR gate and XOR gate are presented as follows.

For OR gate (a, b–input, c–output)

$$FR0(c) = (V0(a) \times V0(b))/P0(c) \quad (3.13)$$

$$FR1(c) = (V1(a) + V1(b) - V1(a) \times V1(b))/P1(c) \quad (3.14)$$

$$BR0(a) = V0(c)/P0(a) \quad (3.15)$$

$$BR1(a)|_{c \text{ is not traced}} = (P1(a) \times R1(c) \times V0(b))/P1(a) = R1(c) \times V0(b) \quad (3.16)$$

$$BR1(a)|_{c \text{ is not traced}} = V0(b) \quad (3.17)$$

For XOR gate (a, b–input, c–output)

$$FR1(c) = (V1(a) \times V0(b) + V0(a) \times V1(b))/P1(c) \quad (3.18)$$

$$FR0(c) = (V0(a) \times V0(b) + V1(a) \times V1(b))/P0(c) \quad (3.19)$$

$$BR0(a) = (V0(c) \times V0(b) + V1(c) \times V1(b))/P0(a) \quad (3.20)$$

$$BR1(a) = (V0(c) \times V1(b) + V1(c) \times V0(b))/P1(a) \quad (3.21)$$

Finally, for gates with multiple fanouts, backward restoration from individual fanout is also treated as *independent* for the sake of simplicity. That is, the backward restorability for the corresponding output node is obtained by combining the restorabilities of its fanouts.

With the our newly-introduced restorability definitions and calculation methods, for the same example shown in [35], the calculated total visibility when tracing FF

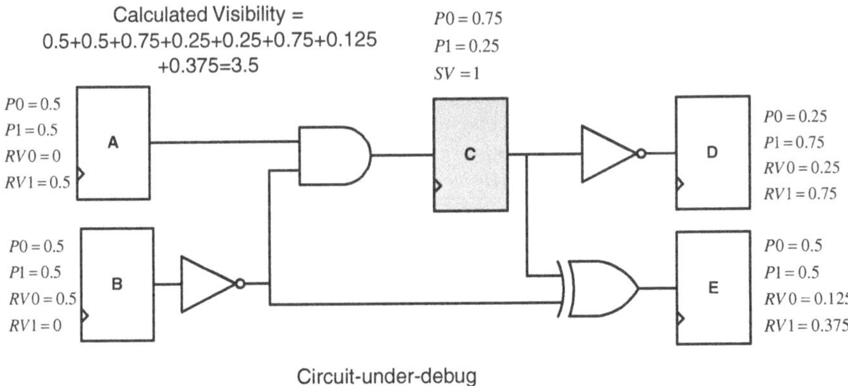

Fig. 3.3 Example with proposed method

C is 3.5 (see Fig. 3.3), which is the same as the actual restoration ratio obtained from simulation for this particular case.

3.3 Trace Signal Selection

Based on the definitions in Sect. 3.2, the automated trace signal selection problem studied in this chapter becomes: *how to set SV to be '1' for a constrained number of signals (FFs and/or input signals), so that the circuit's total visibilities* ($TV = \sum V0 + \sum V1$) *for all state elements is maximized.* To address this problem, this section presents how to calculate the circuit-level visibilities for trace signals and how to use them as guidance to conduct various trace signal selection methods.

3.3.1 Circuit Level Visibility Calculation

Our circuit-level visibility calculation procedure is shown in Fig. 3.4. Starting from the selected trace nodes, we iteratively use forward propagation and backward justification on combinational logic gates (depicted as *search-list* in Fig. 3.4) to obtain the visibilities on other state elements, until TV converges (i.e., no more visibilities can be obtained). For the forward propagation process, we continuously apply the *breath-first* gate-level visibility calculation (see Sect. 3.2) for the following gates connected to the traced signals (denoted as *child node* in Fig. 3.4) to obtain the visibilities for the FFs on the next logic level. The backward justification process is similar. The only difference is that it is implemented in the opposite direction, i.e., update visibilities for the driving gates to the trace signals (denoted as *parent node* in Fig. 3.4) to calculate the visibilities for the FFs on the previous logic level.

Algorithm for circuit level visibility calculation

INPUT: circuit, selected nodes
OUTPUT: *TV*: total visibility

```
Set SV0, SV1 for selected nodes to be "1";
while (TV is not converged) {
   do {
      Put child nodes of FF_i with V_i! = 0 into search-list
      while (Next level FF is not reached) {
         for each node in search-list {
            if (cur_node is passed by backward justification)
               Continue;
            Calculate visibility forward for cur_node;
            Mark cur_node to be covered by forward propagation;
            if (Visibility on child node of cur_node != 0)
               Put the child node of cur_node in search-list;
         }
      }
      Update TV by forward propagation;
   }while (TV is not converged);
   do {
      Put parent nodes of FF_i with V_i! = 0 into search-list
      while (Previous level FF is not reached) {
         for each node in search-list {
            if (cur_node is passed by forward propagation)
               Continue;
            Calculate visibility backward for cur_node;
            Mark cur_node to be covered by backward justification;
            if (Visibility on parent node of cur_node != 0)
               Put the parent node of cur_node in search-list;
         }
      }
      Update TV by backward justification;
   }while (TV is not converged);
}
return TV;
```

Fig. 3.4 Procedure for circuit level visibility calculation

For those untraced FFs with non-zero $V1/\ V0$ after restoration, they can be utilized to further improve the visibilities of other state elements, as can be seen in Fig. 3.5. However, we have to be careful when applying the forward propagation and backward justification interleavedly for visibility calculation. This is because, if the visibility of a gate's output is obtained through forward propagation, this visibility should be not used to further improve the visibility of its inputs through backward justification. Otherwise, we have erroneously over-estimated visibilities with redundant information. Figure 3.6 shows an example for explanation. Initially suppose we have already obtained some visibility of value '1' on two inputs with forward restoration ($FR1(a) = FR1(b) = 0.5$), then by applying forward visibility calculation, the output c gets non-zero visibility correctly ($FR1(c) = 0.25$). However, if we use

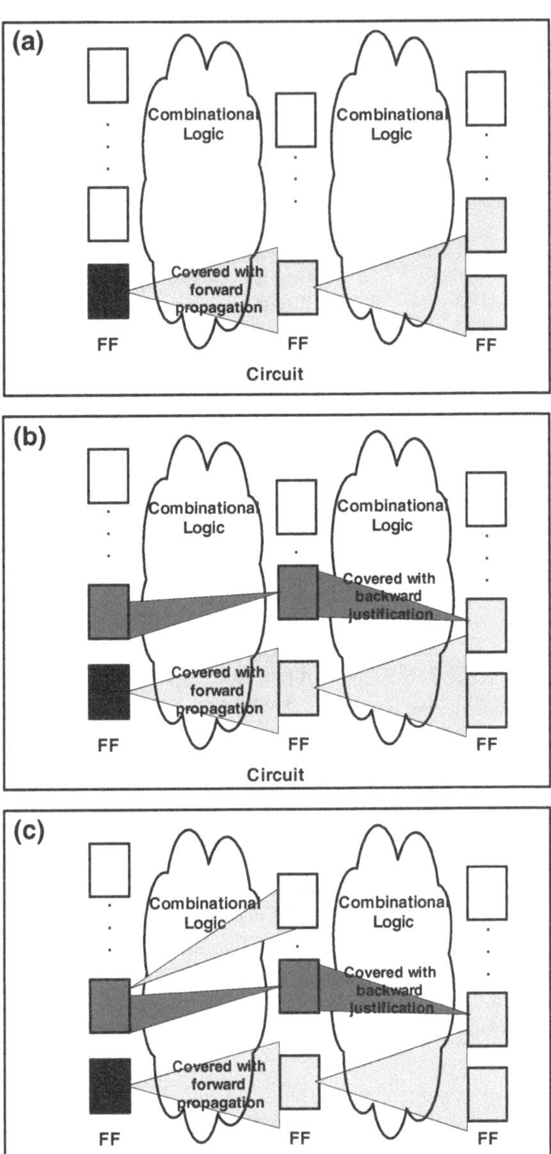

Fig. 3.5 Example of circuit level restoration calculation. **a** 1st forward propagation. **b** 1st backward justification. **c** 2nd forward propagation

the increased value to conduct backward justification, the inputs get increased backward restorability ($BR1(a) = BR1(b) = 0.125$) and the visibilities are increased accordingly (from 0.25 to 0.28125). This kind of calculation mistakenly overestimated the visibility because the output visibility which is solely restored from two inputs can not be utilized again for increasing the input visibilities. Unaware of that and continue to apply calculation iteratively, all relevant visibilities will saturates

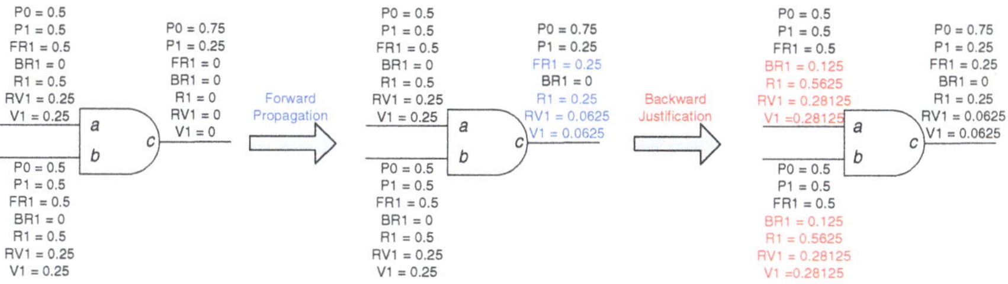

Fig. 3.6 Example of visibility over-estimation with redundant information

to '1', which is obviously over-estimation. To avoid this problem, in our procedure, whenever a gate is utilized for visibility calculation in the forward propagation (backward justification) process, it is labeled so and visibility calculation through it with backward justification (forward propagation) is forbidden.

Similar to [35], the visibility estimation for sequential loops is inherently solved with the above iterative process. The difference from the procedure in [35] is that, they treat forward propagation and backward justification as independent processes, evaluate their restorabilities separately, and then sum them up together. Therefore, for the example in Fig. 3.5 [35], can only calculate the forward restorability in Fig. 3.5a. While for our procedure, we evaluate the impact of the two directions interleavedly, and hence we can further calculate the extended restorabilities as shown in Fig. 3.5b, c.

3.3.2 Trace Signal Selection Methodology

After obtaining the circuit-level visibility for given selected trace nodes using the above procedure, we use them to guide our trace signal selection, which is essentially a greedy heuristic, as shown in Fig. 3.7. In our algorithm, the trace nodes are selected one by one. For each selection, we try every un-selected node and always choose the one that results in the maximum TV together with the already-selected nodes (denoted as maximum cur_TV in Fig. 3.7) with circuit-level visibility calculation. One thing should be noted is that, before the selection process, those nodes that prevent the CUD into functional mode (e.g., reset signal) should be identified, and their effects should be blocked accordingly, because silicon debug should be performed in functional mode.

The computational effort of the above incremental solution can be quite high for large circuit. To address this problem, we propose a multiple signal selection method. In this procedure, we keep a candidate list that records multiple signals providing the maximum TV and select these signals together in each iteration. As shown in Fig. 3.8, during one iteration, every time the cur_TV is obtained for an unselected node cur_node, it is compared with the TV of $last\_node$ in candidate list, which is also the node with smallest TV in the list. Then, if cur_TV is larger, current

Algorithm of greedy signal selection

INPUT: : circuit, the number of selected nodes
OUTPUT: *TV*: total visibility, *SNL*: selected node list

```
Identify nodes that prevent the CUD into functional mode (e.g., RESET);
Block the effect for those nodes;
while (more node can be selected) {
   for each un-selected node {
     Set SV0, SV1 for cur_node and selected nodes to be "1";
     Calculate circuit level visibility;
     if (the returned cur_TV for cur_node is maximum) {
        Record cur_node & cur_TV;
     }
     Update the recorded node as selected one;
   }
 }
return TV, SNL;
```

Fig. 3.7 Procedure for greedy trace signal selection

Algorithm of multiple signal selection

INPUT: : circuit, the number of selected nodes
OUTPUT: *TV*: total visibility, *SNL*: selected node list

```
Identify nodes that prevent the CUD into functional mode (e.g., RESET);
Block the effect for those nodes;
while (more node can be selected) {
   for each un-selected node {
     Set SV0, SV1 for cur_node and selected nodes to be "1";
     Calculate circuit level visibility;
     if (the returned cur_TV for cur_node is larger than the TV of last_node in candidate list) {
        Record cur_node & cur_TV to the end of candidate list;
        Sort the candidate list by TV;
     }
     Update the nodes in candidate list as selected ones;
   }
 }
return TV, SNL;
```

Fig. 3.8 Procedure for multiple trace signal selection

signal *cur_node* replaces the *last_node* and *cur_TV* is also recorded in the list. The candidate list will then be sorted by the decreasing order of TV. After one iteration of visibility calculation for all the un-selected nodes, the remaining nodes in candidate list will be selected together. The above process is conducted iteratively until no more signals can be selected. Consequently, let M be the number of selected nodes for each iteration in above process, then the speedup of the proposed algorithm is roughly M times of the previous one in Fig. 3.7. To note, the selection step M is determined by designers by trading off the selection effort and solution quality.

To further address the running time problem, we can rely on a divide and conquer strategy. To be specific, we can divide the CUD into several regions (e.g., functional modules) and then conduct the proposed solution to select the signals for each region separately. The signals are finally combined together as the traced signals.

3.3.3 Trace Signal Selection Enhancements

3.3.3.1 Trace Signal Selection with "Golden Vector" from High-Level Simulation

In the real application flow of silicon debug, after selected signal data is traced and restored with proposed solution, it will be compared with "golden vector" to detect the erroneous state. After that, starting from the erroneous state, designers can further rely on the logic diagnosis solutions (e.g. [15]) to localize and resolve the root-caused design error. Hence, it is quite beneficial to obtain the "golden vector" providing cycle-accurate and full view of the CUD. However, this type of "golden vector" requires expensive gate-level simulation, and it is not possibly affordable in practical case. As a result, the "golden vector" is usually generated from time-efficient high-level simulation, and it can only provide partial view of the CUD, which means the state values on intermediate signals will not be missing in the "golden vector". In addition, for other state signals the value may not be known for every cycle.

To tackle above problem, our proposed signal selection method needs to be enhanced to work with such "golden vector" from high-level simulation. It is mainly achieved by introducing new evaluation metric to guide the selection procedure. Specifically, we introduce the Golden Vector Weight (GVW) for each to-be-selected signal, which ranges from 0 to 1 and represents the normalized visible weight for each signal in "golden vector". Hence, the GVW of intermediate signal will be assigned with 0, while GVW is 1 for the signals whose value is known for every cycle in "golden vector". For other signals whose value is known for partial cycles, the GVW will be obtained statistically from "golden vector" or pre-defined by designers. Based on above, the evaluation metric for guiding selection process is as follows.

$$TV_{GV} = \sum_{i=1}^{n} GVW_i \times (V0_i + V1_i) \tag{3.22}$$

where n is the number of to-be-selected signals.

By replacing the TV with the newly introduced TV_{GV} in our previously introduced signal selection algorithm (as described in Fig. 3.7), the solution is able to select trace signals providing maximum visibility when working with the "golden vector".

3.3.3.2 Trace Signal Selection with Assertion-Based Debug

Assertion-based verification is quickly adopted in practice for performing hardware verification. In addition [10], introduces the assertion-based solution into silicon debug. To be specific, they insert low-cost assertion checkers (hardware implementation of pre-defined assertions) into the CUD. Then during silicon debug process, when any assertion is violated, the checker will report information on where and when assertions fail, which is an important aid in the debugging process.

To apply our signal selection solution on assertion-based debug, we need to develop novel evaluation metric as well as a new signal selection algorithm. Generally, the assertion denotes the property that the CUD always holds, and it can be represented by the combination of relevant signals, hence the assertion visibility (AV) is defined as.

$$AV = \prod_{i=1}^{n}(A1_i \times V1_i + A0_i \times V0_i) \tag{3.23}$$

in which n is the number of relevant signal values in the assertion, $A1_i = 1$, $A0_i = 0$ if signal i is '1' in assertion, otherwise $A1_i = 0$, $A0_i = 1$. The reason behind the above is that we need to have capability on observing all relevant signal values to monitor an assertion.

Based on the definitions, the trace signal selection problem becomes: how to set SV to be '1' for a constrained number of signals, so that the CUD's total assertion visibility ($TAV = \sum AV$) for all assertions is maximized.

As shown in Fig. 3.9, our algorithm selects trace signals incrementally, and each time we will select the signals relevant to one assertion. For each iteration, we temporarily select the signals within every un-selected assertion signal group, and then

```
Algorithm of signal selection with assertion-based debug
INPUT: circuit, the number of selected nodes, assertion signal groups
OUTPUT: TAV: total assertion visibility, SNL: selected node list

Identify nodes that prevent the CUD into functional mode (e.g., RESET);
Block the effect for those nodes;
while (more node can be selected) {
   for each un-selected assertion signal group {
     Set SV0, SV1 for nodes in cur_group and selected nodes to be "1";
     Calculate circuit level visibility;
     if (the returned ΔTAV/ΔN for cur_group is maximum) {
        Record nodes in cur_group & ΔTAV/ΔN;
     }
     Update the recorded nodes as selected ones;
   }
}
return TAV, SNL;
```

Fig. 3.9 Procedure for trace signal selection with assertion-based debug

together with the already selected nodes, we calculate circuit-level visibility to obtain the total assertion visibility increase ΔTAV. As the associated increase of trace signals ΔN is different, here we will use the average gain $\Delta TAV/\Delta N$ as the normalized evaluation metric and choose the signals in one assertion signal group that provide the maximum average gain. The above procedure is conducted iteratively until no more signals can be selected. Note that, if designers want to observe a large amount of widely-distributed assertions in the CUD, they have to trace more signals at the cost of more DfD hardware.

3.4 Experimental Results

3.4.1 Experiment Setup

We conduct experiments on ISCAS'89 benchmark circuits and compare against [35] to evaluate the effectiveness of the proposed solution (the authors in [35] provided their new results for s38584, s38417 and s35932). As in [35], the trace buffer is defined as $8 \times 4k$, $16 \times 4k$ and $32 \times 4k$ in our experiments.

An event-driven simulator that is able to restore missing states is developed for experiments. As described in Fig. 3.10, compared to traditional simulator, it conducts simulation in both forward and backward directions. Let us use AND gate again to describe how it works in gate-level backward simulation. If the output is logic '1', all the inputs of the gate are set as logic '1'. If the output is '0', only when all the other inputs are known to be logic '1', an input can be determined as logic '0'; otherwise, it is set as a 'X' bit. Known states of traced nodes, dumped from a commercial functional simulator with random input patterns (RESET signal is de-asserted), are used as inputs to our simulator (See Line 1 of Fig. 3.10), while missing states on

Algorithm for event-driven restoration simulation

INPUT: circuit, traced states on selected nodes
OUTPUT: visible states

```
1.  Dump traced states on selected nodes;
2.  while (total visible state number is not converged) {
3.      do {
4.          Event-driven forward restoration;
5.      }while (total visible state number is not converged);
6.      do {
7.          Event-driven backward restoration;
8.      }while (total visible state number is not converged);
9.  }
10. return visible states;
```

Fig. 3.10 Procedure for event-driven restoration simulation

the other nodes are initialized as 'X'. The whole circuit-level simulation flow is also implemented as an iterative process. For each iteration, forward restoration is conducted first and it begins from the first cycle to the last one, and the iterative process stops when no more states can be restored forward. Next, backward restoration is applied in the reverse order, which is also an iterative process and terminates when no more states can be restored backward. To accelerate the simulation process, in both directions of restorations the simulator is implemented as an event-driven one, which means only newly updated states will be further simulated. It should be emphasized that the number of cycles is more than $4k$ in our simulation, because extra states beyond $4k$ cycles are likely to be restored with the sampled signals (e.g., as Cycle 4 in Fig. 3.1b). The simulator ends when no more missing states can be restored.

The simulator is supplied with ten sets of random input patterns and we record the average visibility value in our experiments.

3.4.2 Experimental Results

Tables 3.3, 3.4, and 3.5 present our experimental results when 8, 16, and 32 signals are selected to trace, respectively. Column 2 presents the number of FFs in each

Table 3.3 Experimental result for ISCAS'89 (8 nodes)

Name	# of FF	[35] # of VN	[35] Amount of VS	[35] Restorability ratio	Proposed method # of VN	Proposed method Amount of VS	Proposed method Restorability ratio	Δ (%)	Time (s)
s5378	141	–	–	–	141	480858	14.675	–	5.313
s9234	211	–	–	–	77	156189	4.767	–	9.609
s15850	534	–	–	–	165	652622	19.930	–	104.047
s38584	1426	60	339682	10.366	97	630405	19.238	85.6	162.375
s38417	1636	156	637147	19.444	212	610315	18.625	−4.21	1179.406
s35932	1728	800	2256213	68.854	1344	3546179	108.221	57.8	1099.766

Table 3.4 Experimental result for ISCAS'89 (16 nodes)

Name	# of FF	[35] # of VN	[35] Amount of VS	[35] Restorability ratio	Proposed method # of VN	Proposed method Amount of VS	Proposed method Restorability ratio	Δ (%)	Time (s)
s5378	141	–	–	–	141	589602	8.996	–	13.250
s9234	211	–	–	–	118	470714	7.182	–	24.688
s15850	534	–	–	–	428	1587340	24.221	–	289.453
s38584	1426	77	429749	6.557	162	914904	13.960	112.9	369.500
s38417	1636	231	721152	11.004	487	1220222	18.619	69.2	2647.141
s35932	1728	1440	4255152	64.928	1556	5471084	83.482	28.6	2484.734

Table 3.5 Experimental result for ISCAS'89 (32 nodes)

Name	# of FF	[35]			Proposed method				
		# of VN	Amount of VS	Restorability ratio	# of VN	Amount of VS	Restorability ratio	Δ (%)	Time (s)
s5378	141	–	–	–	141	619393	4.726	–	28.437
s9234	211	–	–	–	155	612343	4.672	–	63.297
s15850	534	–	–	–	469	1743525	13.302	–	692.703
s38584	1426	91	487093	3.716	221	1137593	8.679	133.6	1216.688
s38417	1636	303	949210	7.242	652	1862074	14.206	96.2	5808.235
s35932	1728	1440	4702320	35.876	1721	6105838	46.584	29.8	5217.828

circuit. The restoration results are described in Column 3 to Column 5 and Column 6 to Column 8, for the method in [35] and the proposed method, respectively. "# of VN" means the average number of node that is fully/patially restored, while "Amount of VS" illustrates the average amount of visible states. "Restoration ratio" is the restoration ratio as defined in [35]. The "Time(s)" on the last column is the computational time spent for proposed signal selection.

It is obvious that our selection method which based on more accurate probability definitions achieves a much higher restoration ratio compared to [35] It should be noted that the presented result for [35] is the one with higher ratio from different $Threshold$, 0.1 and 0.5 respectively. For s38584 and s38417, after the first selected 8 nodes, our selection method is still able to restore large amount of states, while the method in [35] can seldom select "essential" nodes (i.e., the restoration ratio drops significantly). Similarly, for s39532, our method can still identify nodes that can restore many missing states after selecting 16 nodes, while the signal selected using [35] cannot restore any missing states (see Table 3.5).

The observed selection trend in our method is that the algorithm firstly selects input control node with many fan-outs, then it chooses those "essential" internal FF nodes. This is reasonable because these control signals can access large amount of FFs and they are on the front end of the circuit. We can also observe from Tables 3.3 to 3.5 that, the restoration ratio drops considerably when more signals are traced, due to the fact that most "essential" signals have been selected to trace in the beginning.

We also conduct the experiments on ISCAS'89 circuits on proposed multiple signal selection that can further accelerate the selection process. As shown in Table 3.6, the total selected trace signal number is decided as 32, and we vary the selection step to be 2, 4, 8, 16 and 32 during different process. Also, the "Ratio" in the table stands for the restoration ratio obtained from corresponding selection and "Time (s)" denotes the execution time. In terms of execution time, As can be observed from the table, in most cases the speedup is more than $2X$ when we double the selection signal step (e.g., from 4 to 8). This is because, the selection process tends to choose the signals with lower visibility propagation capability (in another word, with lower restoration capability) when more signals are selected together during one evaluation iteration. In that case, the effort of visibility evaluation for the selection process

Table 3.6 Experimental result of multiple selection for ISCAS'89 (32 nodes)

	Step=2		Step=4		Step=8		Step=16		Step=32	
Name	Ratio	Time (s)	Ratio	Time (s)	Ratio	Time (s)	Ratio	Time (s)	Ratio	Time (s)
s5378	4.984	19.453	5.027	9.016	4.781	3.985	3.79	1.297	2.494	0.234
s9234	3.481	43.219	3.742	19.5	3.878	8.672	3.289	3.031	3.174	0.39
s15850	8.457	516.562	**10.189**	240.578	10.159	98.313	7.134	34.406	9.297	2.39
s38584	8.194	934.187	7.001	399.672	6.029	176.797	5.514	72.641	4.811	24.375
s38417	10.716	4130.547	9.76	2972.922	9.74	1511.797	7.272	537.828	3.425	64.938
s35932	38.250	4001.109	38.250	1864.906	38.250	815.125	38.250	288.094	38.250	27.891

with larger step will be lower compared to the method with less step number, hence the execution time during each iteration of visibility evaluation is actually less than expected. However, this trend does not hold for doubling the selection step from 1 to 2. As shown in Tables 3.5 and 3.6, the acceleration ratio is less than 2 (e.g., 1.4 for s38417). This is because, when the step difference of two selection methods is small (e.g., 1 for this case), the selection still follows the original greedy selection process to be able to select those essential signals with higher visibility propagation capability. In that case, the effort of visibility evaluation process tends to increase. That is to say, the circuit-level visibility calculation starting from two selected signals will be more than the same process with one signal from the two. Then the acceleration ratio will be lower than 2.

We then study the impact of multiple selection method on restoration ratio. As described in Table 3.6, the general trend is with larger step in selection process, the restoration ratio will be decreased steadily. This is because the process is able to select better signals with more accurate visibility information provided by more effort on circuit-level visibility calculation. In that case, there is a trade-off between selection efficiency and effectiveness. However, this trend does not hold for all cases. For s15850, the ratio is increased from 8.457 to 10.189 with selection step 2 and 4 respectively. We attribute the rarely happened phenomenon to the fact that the selection process is essentially a heuristic and the proposed visibility estimation is not totally accurate because of several simplified assumptions. In that case, it is possible that the selection procedure with less effort on visibility evaluation can be more effective on restoration capability. In addition, we observe that for s35932, the restoration ratio remains to be constant with different selection step. This is because the structure of this circuit is uniform and only a few signals are with much higher restoration capability while for others the visibility is seldom propagated. Then in all selection processes, our method will include these essential signals and achieve the same restoration ratio.

The effectiveness of our proposed method based on probability-based visibility is also verified by the extreme case that we only calculate the visibility for each signal once and select the top ones with highest values, as shown in "Step=32" in Table 3.6. With the process consuming 10 s execution time, we are already able to select the signals to achieve 2.494 to 38.250 restoration ratio.

Table 3.7 Experimental result of selection with assertion-based debug for ISCAS'89 (32 nodes)

Name	# of VA (Pro.)	# of VA (Pro.-A.)	Time (s)
s5378	31964	22951	2.96
s9234	22843	15389	5.28
s15850	18877	21711	5.83
s38584	1260	9335	27.1
s38417	68	18203	34.5
s35932	18564	18278	26.1

Our next experiment is conducted on ISCAS'89 circuits to evaluate the effectiveness of proposed signal selection solutions on assertion-based debug. Here we randomly generate 100 assertions for each circuit, and then follow previous experiment flow to obtain the average number of visible assertions based on traced and restored state value from different selection solution. As described in Table 3.7, Column 2 and Column 3 denote the number of visible assertions with signals selected by the proposed method for maximizing visibility of the CUD (Described in Sect. 3.3.2), and the enhanced solution for maximizing visibility of assertions (Described in Sect. 3.3.3.2), respectively. "Time (s)" on Column 4 is the execution time for the enhanced selection solution targeting assertion-based debug.

We first observe that for small circuits s5378 and s9234, the number of visible assertions by the previous proposed method for maximizing visibility of the CUD is larger than the enhanced one developed for monitoring assertion. This is because the previous method can restore large part of state value within the small circuits by tracing selected 32 signals, so that the number of visible assertions, which are the combination of visible states, will also be large. Meanwhile, the enhanced solution constrains the to-be-selected signal within the assertions and it may miss the signal outside of the assertions, but with high restorability. The phenomenon also holds for large circuit s35932, as the restorability is extremely high with previous method and most part of state value becomes visible. Above demonstrates the effectiveness of our previous proposed method for working with assertion-based debug, as long as the traced and restored state value will cover most part of the CUD. However, we can also see that for other cases of large circuits, the previous solution will achieve much less visible assertions than the enhanced solution. For the circuit s38417, the visible assertions by previous method is only 68, compared with 18203 from enhanced solution. We attribute phenomenon to the fact that the previous method will restore relatively small part of state value for large circuit by tracing only 32 signals, so that it will easily miss the assertions outside, and for this kind of case, the enhanced solution developed for monitoring assertion will achieve satisfying visibility on targeted assertions. On the other hand, as the enhanced solution only takes 10s to select signals for monitoring assertions for all ISCAS'89 circuits, it demonstrates that the solution is able to work efficiently with assertion-based debug on large industrial circuit.

3.5 Conclusion

In this chapter, we propose the methodology that automatically selects signals to be traced in real time in post-silicon validation to achieve maximal circuit visibility. The method is from the foundation of probability-based *visibility* definition and corresponding calculation method. We also developed the multiple signal selection to accelerate the process. In addition, we propose the new metrics to guide signal selection for effectively working with high-level simulation and assertion-based debug, respectively. Experimental results demonstrate the efficacy of our proposed signal selection technique which is superior to existing solutions.

Chapter 4
Multiplexed Tracing for Design Error

Besides the trace signal selection problem that has been explored in Chap. 3, existing trace-based solutions have another limitation. They typically trace the same set of selected signals throughout each debug run. It might not be a problem for applying trace-based solution on dedicated purposes, e.g., to reconstruct instruction flow [53] or to monitor bus communication protocol [10]. When debugging general logic circuits, however, this kind of "static" signal tracing methods limits the visibility of the design since it only provides limited part-view of the CUD, while the states of other tapped signals are not visible. Consequently, if the error effects do not manifest themselves on those signals that are currently under trace, more debug runs need to be conducted, which greatly increases the bug localization effort.

Intuitively, for a permanent design error that has been activated, if the trigger mechanism is set properly and its effect can be propagated to one or more tapped signal, there should be a high possibility to catch it when the number of trace cycles is sufficiently long and further increase of trace cycles does not improve the debuggability much. Based on the above, in this chapter, we propose a multiplexed signal tracing strategy that is able to significantly improve the debuggability of the circuit, by leveraging the fact that the number of tapped signals is much more than the number of signals that can be traced concurrently. That is, we divide the tracing procedure in each debug run into a few periods and intelligently trace a different subset of accessible signals (i.e., tapped signals) in each period.

The remainder of this chapter is organized as follows. Section 4.1 reviews related works and presents the motivation of our work. In Sect. 4.2, we define the proposed metric used to evaluate the effectiveness for capturing the evidence of design error. In Sect. 4.3, we describe the proposed methodology in detail. Experimental results on benchmark circuits are shown in Sect. 4.4. Finally, Sect. 4.5 concludes this work.

X. Liu and Q. Xu, *Trace-Based Post-Silicon Validation for VLSI Circuits*,
Lecture Notes in Electrical Engineering, DOI: 10.1007/978-3-319-00533-1_4,

4.1 Preliminaries and Summary of Contributions

Design errors are caused by erroneous design process and they violate the pre-defined specification of IC products. Due to the ever-increasing complexity of IC designs, there is a growing number of such errors left in the first silicon, requiring post-silicon debug to catch them [58]. As shown in Fig. 4.1, during debug phase, certain test input vectors fed into the fabricated chip can activate the bug so that its error effects manifest themselves. Since the circuit under debug is a piece of silicon that has already been fabricated, the main challenge here is that bugs may take long time to be activated and there is only limited visibility of its internal signals. Consequently, an essential step is to capture the error evidences within the chip, so that designers can quickly focus on a small region of the CUD, and then apply various diagnosis methods (e.g., [15]) to root-cause the bug and fix it.

One widely-used technique to mitigate this problem is to reuse the CUD's existing test structure (e.g., scan chains) to run/stop its operation and observe whether the values in the circuit's storage elements are the expected values [31, 59]. However, this low-cost technique provides little help for tracking those tricky bugs that takes a long period of operation to manifest themselves. In addition, it is a challenging problem to repeat the CUD's error occurrence procedure in a deterministic cycle-accurate manner.

To tackle the above problems, trace-based post-silicon debug solutions are proposed to observe the states of selected internal signals in the CUD at real-time. Figure 2.1 depicts a conceptual hardware infrastructure for trace-based debug techniques. As signal tracing involves non-trivial DfD overhead, only a small portion of "key" signals in the circuit are tapped, and in each debug run, a subset of the tapped signals are traced concurrently. An interconnection fabric is used to link the large number of tapped signals to the trace buffers/ports. The trigger unit controls the start

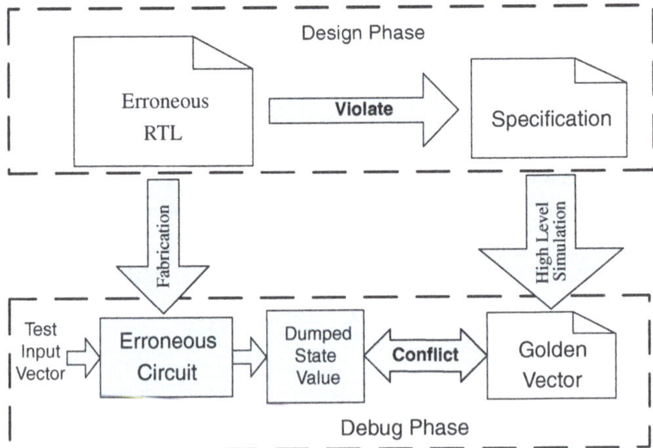

Fig. 4.1 Post-silicon debug for design errors

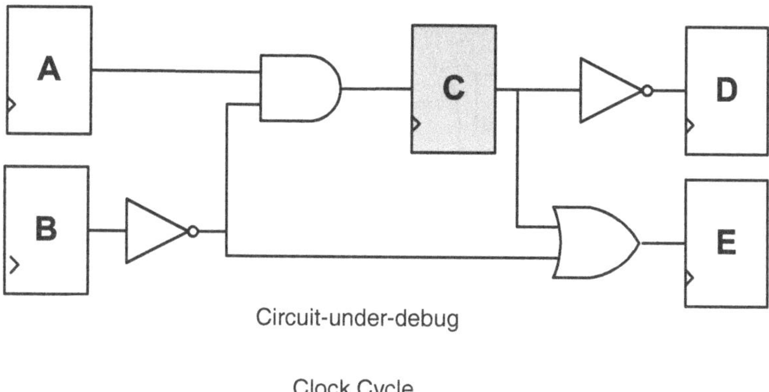

Clock Cycle

	0	1	2	3	4
A	1	1	X	X	X
B	0	0	X	X	X
C	0	1	1	0	X
D	X	1	0	0	1
E	X	X	1	1	X

Restored data in flip flops

$$Restoration\ Ratio = \frac{14}{4} = 3.5$$

Fig. 4.2 State restoration example

and stop of the tracing, in which the triggering mechanism can be configured through JTAG interface through the debug configuration channel [68].

Selecting which signals to tap in the design is a key issue for effective debug. To maximize the visibility of the CUD, some researchers proposed to select signals based on the "state restoration" concept [35, 42, 56]. An example is shown in Fig. 4.2. Since some states can be restored with logic implication from traced FF(C), 14 states are restored from 4 traced ones. Using such restoration capability as evaluation metrics to guide trace signal selection, however, may miss finding certain bugs. Consider the example in Fig. 4.2 again. Suppose the OR gate is mistakenly replaced by another gate (due to some design errors), it is only beneficial if we trace FF(E) to capture the evidence induced by this error. We cannot identify this error by tracing other FFs (e.g., FF(C)) and applying *golden* netlist-based logic implication to root cause it, even if the corresponding restoration ratio is high. In [78], the authors proposed a different trace signal selection method, targeting on minimizing the latency for capturing the propagated evidences after bugs are triggered. The method, however, is based on the assumption that the states' sequence for exposing error is available, which may not be obtained as early as the design phase. Furthermore, another key

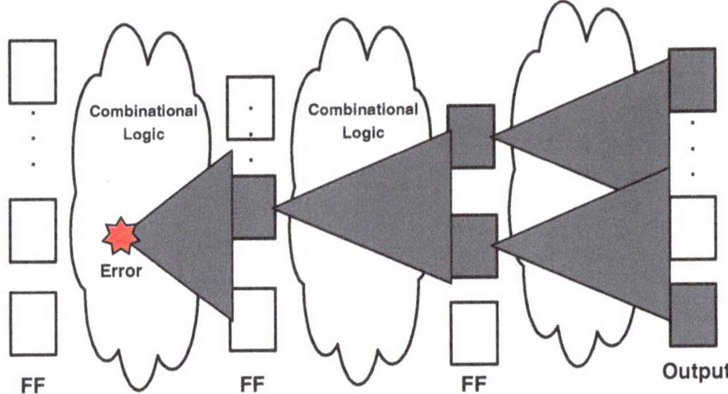

Fig. 4.3 Example of design error evidence propagation

issue of debuggability on how to determine a subset of accessible signals to trace in the debugging phase remains untouched.

Tricky design errors may take millions of cycles to be activated during debug runs [14], when the CUD is accidentally operated into "corner cases". If the trigger mechanism is properly designed, however, the error can generate its evidences from time to time during the running procedure. In addition, as demonstrated in Fig. 4.3, the evidences will further propagate forwardly and leave their tracks on FFs (e.g., the FFs in bold in Fig. 4.3). This evidences can also be masked by the controlling values of side-inputs in the process of going through a logic gate. As a result, some FFs cannot capture evidences (e.g., the FFs in blank in Fig. 4.3). With this observation, if we are able to observe the error evidences by properly tracing relevant FFs for some time, the region with error is greatly zoomed in, and hence the root-cause effort is significantly reduced. Besides, the tracing time is not necessary to be as many as possible. In [56], the authors proposed to select and trace two sets of signals in odd time-frames and even time-frames respectively. Such multiplexed tracing method, however, requires interconnection fabric to redirect debug data flow in every clock cycle, which involves large DfD overhead and high power consumption.

In trace-based debug solutions, the number of accessible signals is much more than the number of signals that can be traced concurrently in each cycle. Hence, for a suspicious region that is likely containing errors, it is possible to divide the tracing procedure into different periods and intelligently select a group of "key" signals from accessible ones for tracing in each period. This strategy can enhance the capability of capturing error evidences for the region. Motivated by above, we propose the multiplexed signal tracing method. The main contributions include

- we define *design error visibility* (*DEV*) to measure the tracing quality;
- we propose novel automated trace signal grouping algorithms that are much more effective than existing "static" tracing techniques and multiplexed tracing with randomly-grouped trace signals;
- we design supporting hardware with negligible cost.

Table 4.1 Terminologies for visibility calculation

Selected visibility ($SV1/SV0$)	For selected trace nodes, $SV1 = SV0 = 1$; otherwise $SV1 = SV0 = 0$
Design probability ($P1/P0$)	The probability that the node is '1'/'0' in functional mode
Evidence impact ($EI1/EI0$)	The probability that evidence '1'/'0' is propagated on one FF
Evidence visibility ($EV1/EV0$)	The probability that the evidence '1'/'0' is visible to be captured on one traced FF
Design error visibility ($DEV1/DEV0$)	The probability that the design error with evidence '1'/'0' is visible with traced signals
Total design error visibility ($TDEV$)	The capability that design errors in the suspicious region are visible with traced signals

4.2 Design Error Visibility Metric

As discussed in Sect. 4.1, it is essential to introduce a reasonable metric to evaluate the effectiveness on capturing the evidences from possible design errors within a suspicious region. We start with emulating the actual behavior of error evidence propagation, as demonstrated in Fig. 4.3. To be specific, when a design error is activated in functional mode, the phenomenon that a correct '0' is replaced by an erroneous '1' or a correct '1' is replaced by an erroneous '0' occurs on at least one node of the circuit. Then the erroneous value propagates in the circuit and might be captured by FFs or outputs. Based on this observation, we first introduce *Evidence Impact* (denoted by $EI1/EI0$) to represent the probability that the evidence is propagated to a certain FF (see Table 4.1). Different from similar concept of fault detection probability in manufacturing test [29], for design error we cannot predict the error occurrence probability, hence Evidence Impact is set to '1' at the possible root-cause location. With the forward propagation, the evidence tends to be masked because some other side-inputs can be controlling values. To indicate this weakening effect, we then introduce a series of *weakening parameters (WP)* for various types of gates and express them as the following equations.

$$WP_{\text{and/nand}} = \prod_{i=1}^{n_{\text{input}}-1} P_i(1) \tag{4.1}$$

$$WP_{\text{or/nor}} = \prod_{i=1}^{n_{\text{input}}-1} P_i(0) \tag{4.2}$$

$$WP_{\text{not/xor/xnor}} = 1 \tag{4.3}$$

where, n_{input} is the number of inputs to the gate and $P_i(0/1)$ is the probability of the logic value (0/1) on side-input i of the gate. Several methods can be utilized to obtain the above probability. One method is to run simulation with operational input sequence and then dump internal values to estimate it. Alternatively, we can simply use structural analysis to calculate the probability forwardly from primary inputs, by assuming some control inputs are pre-set as 0/1 to ensure the CUD is working in functional mode, while all other inputs are with the probability 0.5 to be value 0/1.

With these notations, after the evidence passes through a gate, it is weakened as $EI_{out} = EI_{in} \times WP$. Re-convergent fan-out may cause multiple propagated evidences to propagate through the same gate. To capture this effect, we introduce the expression $EI_{out} = \bigcup EI_{in_{evi.}} \times WP_{non-evi.}$, within which the $WP_{non-evi.}$ is determined by the inputs without propagated evidences.

If there is a chance that the evidence is captured by a certain FF, we need to trace this FF such that the event that the evidence reaches the FF can be observed. We therefore need to define the selected visibility (SV) to represent the monitor capability of the traced signals (the detailed definition are shown in Table 4.1).

Sequentially, we use Evidence Visibility (EV) to indicate the probability that the evidence is visible to be captured on a under-traced FF, which is given by

$$EV1 = SV1 \times EI1 \tag{4.4}$$

$$EV0 = SV0 \times EI0 \tag{4.5}$$

Since the evidences generated from an error can be captured by one or more traced FFs and the observation of any evidence is helpful for detecting the error, we regard the events that each evidence is captured by traced signals as mutually independent. With this assumption, the probability that a possible design error with evidence ‘1’/‘0’ is detected with traced signals can be expressed as

$$DEV1 = \bigcup_{k=1}^{n_{\mathrm{FFcap}}} EV_k 1 \tag{4.6}$$

$$DEV0 = \bigcup_{k=1}^{n_{\mathrm{FFcap}}} EV_k 0 \tag{4.7}$$

where n_{FFcap} is the number of FFs that capture error evidences.

Moreover, in the context of multiplexed tracing, we have several tracing periods and the evidences’ capture in any period is helpful for detecting the corresponding error. To capture this effect, we can rewrite Eq. (4.8)–(4.9) as follows.

$$DEV1 = \bigcup_{j=1}^{m_{\mathrm{period}}} \bigcup_{k=1}^{n_{\mathrm{FFcap}}} EV_k 1 \tag{4.8}$$

$$DEV0 = \bigcup_{j=1}^{m_{\mathrm{period}}} \bigcup_{k=1}^{n_{\mathrm{FFcap}}} EV_k 0 \tag{4.9}$$

where m_{period} is the number of tracing periods. Again, the evidences’ captures between different period are regarded as independent events.

For a possible design error, we can use the summation of its design error visibilities with respect to evidence ‘1’ and ’0’ to evaluate the detection capability for the error. The summation of this quantity over all possible errors on suspicious cells is finally defined as Total Design Error Visibility (TDEV), which indicates the overall capability of detecting design errors in the suspicious region, and can be expressed as

$$TDEV = \sum_{i=1}^{n_{\text{cell}}} (DEV0 + DEV1) \tag{4.10}$$

where n_{cell} is the number of cells where possible errors can occur (namely, suspicious cells) in the suspicious region.

By emulating evidence propagation behavior, this metric provides a reasonable estimation of the propagation impact for each possible error, which facilitates to identify those FFs with high chances to detect errors. At the same time, by combining the visibilities of all suspicious cells (see Eq. (4.8)–(4.10)), we inherently balance the error detection capability for all suspicious cells in signal grouping. To be specific, every time we choose a new trace signal, the benefits of potential trace signals are evaluated by the induced gain on $TDEV$. Thus, generally speaking, the visibilities of the possible errors that already have high DEV tends to result in less $TDEV$ gain with new tracing signal; while the errors with low DEV tend to provide more $TDEV$ enhancement. Consider a simple case as an example. Given two suspicious cells a and b, at the end of the first period we have $DEV1(a) = 0.9$, $DEV1(b) = 0.1$, $DEV0(a) = 0$, and $DEV0(b) = 0$. In the second period, suppose we have two options to get 0.5 $DEV1$ for a and b for this period. Combining the detection capability contributed by the two periods with Eq. (10), the resulting $DEV1(a) = 0.95$ and $DEV1(b) = 0.55$. With Eq. (12), we will select those trace signals that enhance the error detection capability of suspicious cell b because they bring more $TDEV$ gain. From the above, the metric $TDEV$ inherently guides the selection of trace signals to cover possible errors with low detection capability during signal grouping procedure.

4.3 Proposed Methodology

The design flow of the proposed multiplexed signal tracing method is described in Fig. 4.4. During design phase, supporting DfD circuitries to facilitate multiplexed tracing (e.g., interconnection fabric and debug controller as shown in Fig. 2.1) are inserted in the design (detailed in Sect. 4.3.1). Then, during the post-silicon debug phase, for a particular suspicious region relevant to one or more trigger conditions, we use an off-chip algorithm to determine signal grouping for maximizing error detection capability in that region (detailed in Sect. 4.3.2). The arrangement is loaded into on-chip debug controller through JTAG interface to facilitate multiplexed trace control.

When certain pre-defined condition is triggered in a debug run, the debug controller starts to trace data in a multiplexed fashion. The dumped information are then analyzed off-line by designers to root cause design errors. The debug process

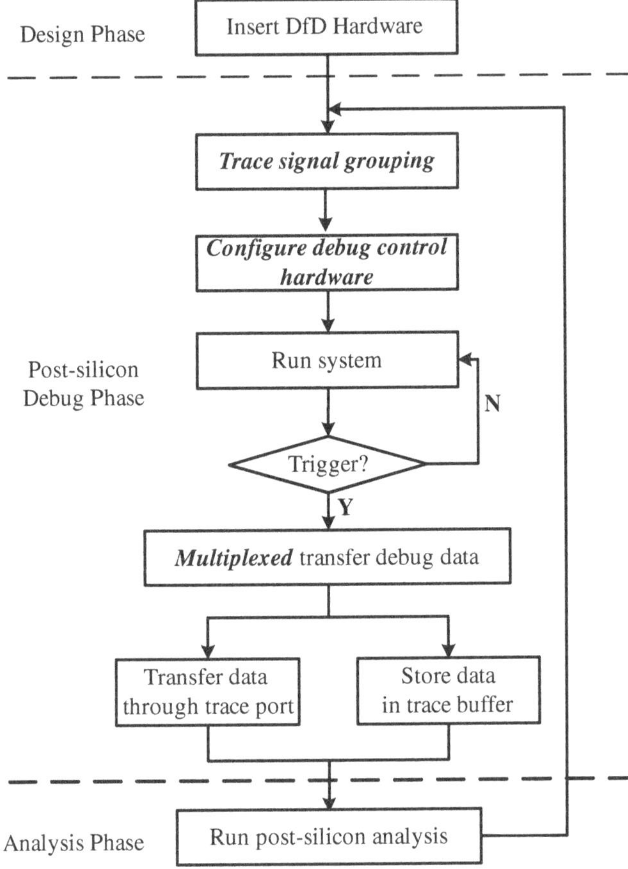

Fig. 4.4 Proposed trace-based debug scenario

is terminated if we can successfully find the bugs. Otherwise, we either try to zoom in the suspicious region with the help of captured evidences or switch to the other part of the CUD when no error evidence is detected during the previous debug run. In both cases, the signal grouping algorithm is used again to obtain the new set of traced signals and the associated configuration is loaded into debug controller for another debug run. The above process is conducted iteratively until all design errors are found and eliminated.

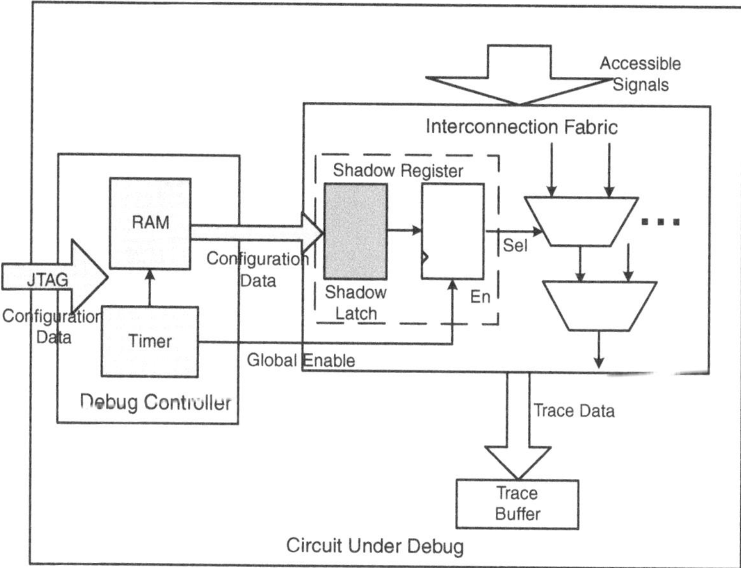

Fig. 4.5 Diagram of supporting DfD hardware

4.3.1 Supporting DfD Hardware for Multiplexed Signal Tracing

To facilitate multiplexed signal tracing that are composed of multiple tracing periods, a few modifications are required on top of the conventional trace-based DfD infrastructure, as indicated in Fig. 4.5.

First of all, *shadow* registers are added into the configuration unit, which determines which signals are transferred through interconnection fabric. Within each tracing period, the shadow latches in these registers can be loaded with the configuration data for next tracing period without intervening the normal trace data transfer. Then, when the new period starts, all the *shadow* registers are updated by a global enable signal from debug controller to configure the trace interconnection fabric simultaneously so as to transfer data from another group of accessible signals to trace buffers/ports. It is important to note that, the time required to load the shadow register determines the minimum number of cycles of each tracing period.

In addition, a small RAM needs to be introduced into the on-chip debug controller to store the configuration data and a timer that controls the data loading into *shadow* registers and assert the global enable signal at the beginning of each tracing period. The controller can be configured through JTAG interface.

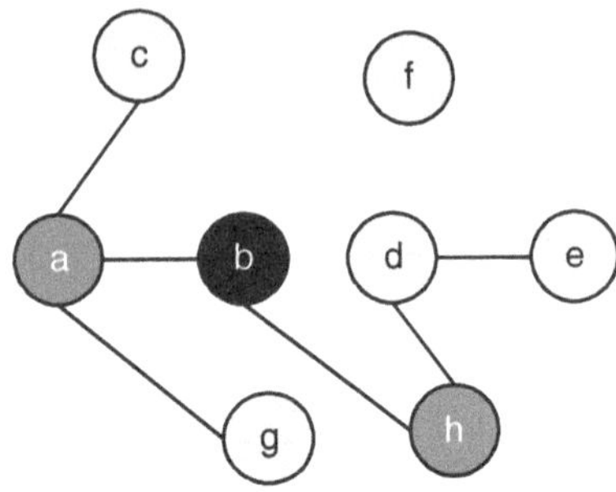

Fig. 4.6 General interconnection fabric: an example

4.3.2 Signal Grouping Algorithm

As not all signals can be traced concurrently due to the existence of the trace interconnection fabric (e.g., any signals going through the same multiplexer cannot be traced concurrently), it is essential to develop an algorithm to judiciously group signals in each tracing period to maximize the error detection capability.

Based on the metric introduced in Sect. 4.2, the above problem is formulated as: Given

- Suspicious region under debug;
- Concurrent tracing constraint from interconnection fabric;
- Relevant accessible signals[1] determined by designers;
- Trace buffer size;
- The number of tracing periods;

To maximize TDEV (i.e., the total design error visibility) of the circuit under debug.

We tackle this problem in two steps. Firstly, with a set of targeted cells in suspicious region, we extract the relevant FFs that have the potential to catch evidences and estimate corresponding evidence impact by following their evidences' propagation tracks. In this step, For each suspicious cell, we initialize $EI0/EI1$ to be 1, which is for error evidence 1 and 0, respectively. We then propagate them forwardly. The impact is weaken during propagation so that the process will be terminated when the impact is close to 0. It can also be stopped by bounding the number of relevant FFs to a pre-defined threshold due to the memory cost, as capturing the propagated error evidences on a few nearby FFs is sufficient to detect the root-cause error. After that, we propose a heuristic signal grouping algorithm to maximize TDEV for the general type of interconnection fabric as follows.

A general trace interconnection fabric can be the widely-adopted MUX network, the one with high flexibility introduced in [41] or any type of design specific structure. We represent the concurrent tracing relationship within the fabric as a graph, as the example shown in Fig. 4.6. Within the graph, a vertex denotes each accessible signal while an edge denotes the two connected signals cannot be traced together (e.g., they

[1] Not all accessible signals are relevant for each debug run.

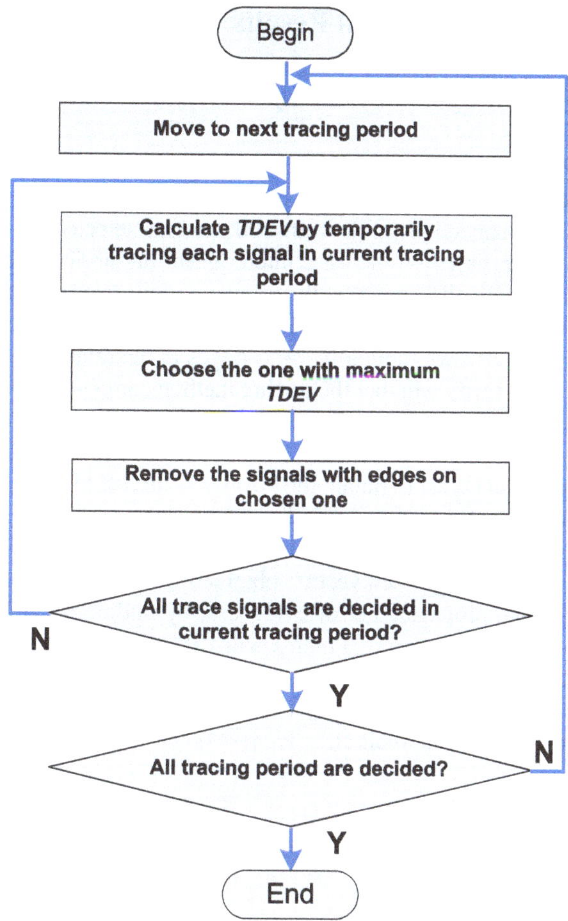

Fig. 4.7 Flowchart of signal grouping with general interconnection fabric

go through the same multiplexer). Hence a unified MUX network can be represented by a few cliques with no edge in between in the graph.

With the above graph representation, we propose a greedy method as depicted in Fig. 4.7 to solve the problem, wherein we incrementally select trace signals period by period to maximize TDEV. To decide which signals to trace in each sample period, we firstly estimate the resulted TDEV by temporarily choosing each relevant accessible signal remaining in the graph. Then, the one with the maximum $TDEV$ is chosen (e.g., b in Fig. 4.6). After that, we remove all nodes connected with the already-chosen one (e.g., a and h in Fig. 4.6) from further selection in the current tracing period, as these signals cannot be traced together with the selected one in this period. The procedure for each sample period ends when the number of selected signals reaches trace buffer width. As discussed in Sect. 4.2, the metric TDEV itself inherently guide us to maximize the probability of detecting design errors within the suspicious region.

4.4 Experimental Results

4.4.1 Experiment Setup

We conduct experiments on ISCAS'89 benchmark circuit s13207 and s38417 to evaluate the effectiveness of the proposed solution, and we consider the general interconnection fabric with 100 random selected accessible signals. As discussed earlier, based on the trace interconnection fabric implemented in the CUD, we can automatically construct the corresponding relation graph (see Fig. 4.6) to represent incompatibility among trace signals. Since this is not available to us, for the sake of simplicity, we randomly insert edges in the relation graph in our experiments.

To verify whether the tracing method can observe the errors' evidences or not, we randomly generate 1000 errors based on the widely-used "cell replacement" error model in the literature [15]. Each time, we inject one of the errors in the circuit into the original netlist to obtain the erroneous netlist. Simulation is then conducted to dump the actual states. As no dedicated trigger mechanism is used in our experiments, the state dumping starts from the beginning of the simulation. These states are compared against the "golden vector" obtained from the simulation with the original netlist to get the propagated error evidences, by finding the difference between actual states and "golden vector". Finally, with different signal tracing methods, the evidence can be treated as visible if the capture FF is currently traced at the particular cycle. We can compare the number of visible errors from various signal tracing methodologies to demonstrate their effectiveness on error detection.

4.4.2 Experimental Results

Figures 4.8 and 4.9 present our experimental results for s38417 and s13207, respectively. We set the length for each tracing period to *1k* (i.e., 1024) clock cycles. We also set the trace buffer size to be *64k* and *128k* in total and vary their width and corresponding depth to obtain the results with different buffer usage strategy. For example, the *64k* buffer can be used as $8 \times 8k$ or $16 \times 4k$. The buffer usage strategies are marked on X-axis. Hence, the left half of the figure shows the results for the *64k* buffer, while the right half is for the *128k* buffer. Buffer depth infers the number of tracing periods, e.g., when the buffer depth is set as *8k*, we have $\frac{8k}{1k} = 8$ tracing periods.

We compare the multiplexed tracing signals grouped with the proposed method (denoted by "Muxed Pro.") with other solutions, including non-multiplexed tracing signals grouped randomly (denoted by "Non-Muxed Rand"), non-multiplexed tracing signals grouped by proposed method (denoted by "Non-Muxed Pro.") and multiplexed tracing signals grouped randomly (denoted by "Muxed Rand"). All of them target the same suspicious region covering all the injected errors, assuming all the accessible signals are relevant. The results of the proposed evaluation metric

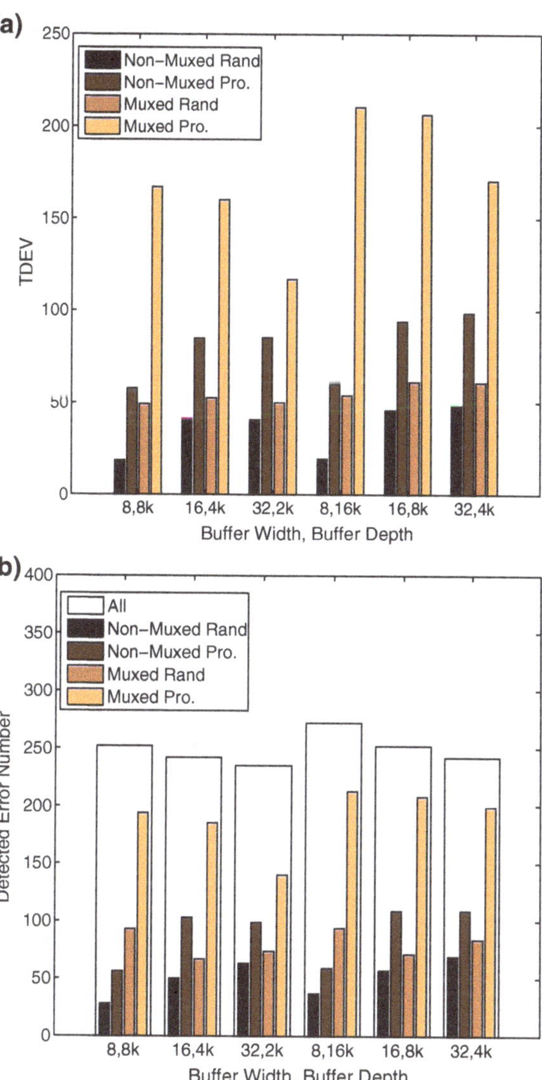

Fig. 4.8 Experimental results for s38417. **a** TDEV comparison. **b** Detected error number comparison

and the number of detected errors resulting in evidences on traced information are denoted by "TDEV" and "Detected Error Number" in these tables, respectively.

First of all, the trend with our proposed evaluation metric $TDEV$ in Figs. 4.8 and 4.9 is roughly consistent with that of the number of detected errors, which demonstrates the effectiveness of this metric. At the same time, there are a few exceptions (e.g., for s13207 when the buffer is configured as $8k \times 16$). This is because, the $TDEV$ is for evaluating the error detection quality for the whole suspicious region, while the errors that are actually caught are just part of the signals in the

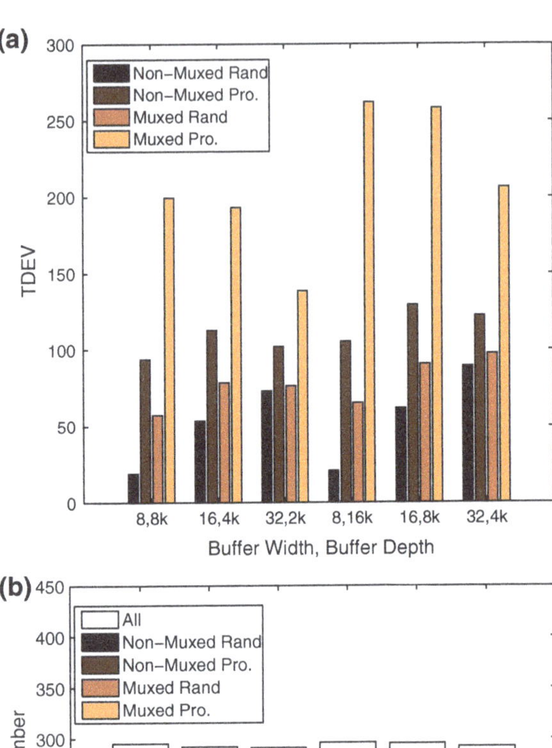

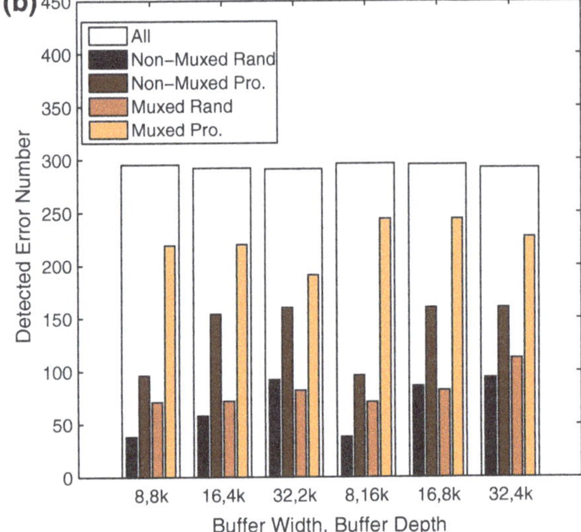

Fig. 4.9 Experimental results for s13207. **a** TDEV comparison. **b** Detected error number comparison

suspicious region and they are also related to the particular debug run, which is unknown during the design stage.

In Figs. 4.8b and 4.9b, the white bars with the legend "All" indicate the number of errors that result in evidences on accessible signals of the CUD. We can observe that only part of the injected errors can result in evidences on the trace-based DfD structure (24.2–29.6 %). This is because: (i). the injected errors can only be activated by certain input sequence, which may not occur during the tracing cycles. (ii). the evidences generated by errors can be masked during the propagation so that they will

Table 4.2 Area overhead of proposed multiplexed tracing

Buffer type	$8 \times 8k$	$16 \times 4k$	$32 \times 2k$	$8 \times 16k$	$16 \times 8k$	$32 \times 4k$
# of Period	8	4	2	16	8	4
$\Delta(\%)$	3.50	1.78	0.90	3.47	1.76	0.89

$\Delta = \frac{Extra\ Hardware\ Cost}{Conventional\ Hardware\ Cost} \times 100\%$

not result in evidences on any FFs. (iii). The trace-based DfD structure only allows to trace 100 out of 1636 and 638 FFs for circuit s38417 and s13207, respectively, and hence those signals that capture error evidences may not be accessible. We also notice that the number of detected errors grows slowly with the increase of trace cycles. This phenomenon can be attributed to the fact that some errors are activated by certain condition that rarely occurs. In this case, the extension of tracing procedure can result in higher error detection chance. To note, the number of detected errors indicated with "All" tends to be higher than the one with the optimal trace signal grouping solution. This is because the DfD structure constraint only a small part from accessible signals to be traced. We can observe from results that our solution by tracing part of accessible signals can detect up to 82.7 % of all detectable errors (as shown in Fig. 4.9b), which demonstrates the effectiveness of our method. We also observe that most evidences captured by accessible signals are within two sequential levels from the root-cause, showing the effectiveness of trace solutions on error localization.

From the results, we can observe that the proposed multiplexed tracing solution outperforms significantly over the non-multiplexed one with the same signal grouping used in our first tracing period. Even for the closest case having two tracing periods, utilizing multiplexed tracing can still detect on average 27.8 % more errors than the other. Even with random grouping, multiplexed tracing detects on average 37.3 % more errors than the non-multiplexed one. When comparing against the case by randomly grouping a set of signals for each tracing period, as shown from Figs. 4.8 and 4.9, on average the proposed one detects 1.75 times more errors than the random solution. Even by comparing the non-multiplexed tracing with signals selected with proposed method against the multiplexed one with random grouping, the first one detects more errors in many cases (e.g., the result in Fig. 4.9b).

The computational time of the proposed method is acceptable. It takes only tens of seconds for all cases on s38417 and less than ten seconds for s13207. For s38417, a large share of the computational effort is spent on error effect estimation, its runtime is almost independent of buffer usage strategy. While for s13207, the runtime almost grows linearly with the number of tracing periods for s13207. This is because it does not take much time for error effect estimation due to the small size of s13207, and hence most effort is spent on signal grouping.

Finally, to evaluate the area cost for the proposed solution, we implement and synthesize the DfD hardware using a commercial tool for multiplexed tracing and non-multiplexed tracing, respectively, with different buffer configurations. As shown in Table 4.2, the additional DfD area overhead (in NAND2 gate equivalent) of the proposed solution is quite small, less than 3.5 % of the DfD cost for conventional non-multiplexed tracing. In particular, we can observe that within the DfD structures to facilitate multiplexed tracing, the storage element in debug controller dominates the extra DfD cost, while most of the hardware cost for conventional tracing is from trace buffer. Hence, for different trace buffer utilization types of the same capacity (e.g., $8 \times 8k$, $16 \times 4k$ and $32 \times 2k$), the original DfD cost remains almost constant, while the size of storage elements in our debug controller decreases linearly by the number of tracing periods (e.g., 8–4). Hence, the hardware overhead decreases correspondingly as shown in Table 4.2. It also explains why we find similar overhead when we double the trace buffer capacity (e.g., $8 \times 8k$ to $8 \times 16k$), as the required storage element in debug controller also doubles its size with the increase of tracing periods (e.g., 8–16). To note, the number of tracing periods can be flexibly determined to tradeoff debuggability and DfD cost.

4.5 Conclusion

In this chapter, we propose a novel multiplexed signal tracing method to maximize the design error detection capability, under various trace interconnection fabric constraints. Experimental results on ISCAS'89 benchmark circuits demonstrate the effectiveness of our solution.

Chapter 5
Tracing for Electrical Error

The duty of post-silicon validation is to eliminate various errors escaped from pre-silicon verification, which can be broadly classified as functional errors and electrical errors. Chap. 3 and Chap. 4 target on resolving functional errors. These errors, being repeatable, are relatively easy to be identified and resolved by designers [59, 70]. Electrical errors, however, only occur in certain electrical environment during normal operation and they may take millions of cycles to expose themselves. Hence, they are extremely challenging to be repeated and root-caused. For example, conducting silicon debug in test environment (using automatic test equipment) may simply result in "No Trouble Found (NTF)" because the CUD's behavior is quite different in system's functional environment. Existing solutions mainly resort to designers' own experiences to identify such errors (e.g., by analyzing voltage-frequency shmoo plot [30]), which cannot guarantee the debug quality and efficiency.

To alleviate the limitation of the above manual process, we propose to monitor and trace those internal signals in the CUD that are related to its electrical abnormal behavior. Such real-time visibility facilitates us to identify the root cause of the electrical errors.

Electrical errors often lead to reduced operational frequency of the CUD, due to parasitic coupling noises between wires, power supply noise, and/or insufficient driving strength, etc. In this chapter, we first model the behavior of such speedpath-related electrical errors. Accordingly, with given DfD constraint, we propose an automated trace signal selection methodology to maximize the error detection probability with a subset of signals relevant to the targeted speedpaths. Moreover, to reduce the storage requirements for tracing, we develop a novel trace qualification technique that employs reconfigurable logic to store useful traced data only when the errors are detected. The proposed technique hence significantly improves the utilization of the trace buffer. To the best of our knowledge, this is the *first* trace-based solution for debugging electrical errors in general logic circuits in post-silicon validation. With the proposed technique, we can detect speedpath-related electrical errors at its root-caused site, on the exact error occurrence cycle, without requiring

X. Liu and Q. Xu, *Trace-Based Post-Silicon Validation for VLSI Circuits*, Lecture Notes in Electrical Engineering, DOI: 10.1007/978-3-319-00533-1_5,

any supporting "golden vector", which are often unavailable when debugging tricky errors.

The remainder of this chapter is organized as follows. Section 5.1 reviews related work and motivates this work. Section 5.2 describes the speedpath-related electrical error model and equations to obtain corresponding visibility. In Sects. 5.3 and 5.4, we illustrate the proposed signal selection and data qualification techniques in detail. Experimental results on benchmark circuits are then shown in Sect. 5.5. Finally, Section 5.6 concludes this work.

5.1 Preliminaries and Summary of Contributions

Trace-based debug solution facilities designers to observe abnormal behavior of circuits in operational mode and conduct root-cause analysis, and hence has been widely adopted in the industry (e.g., [1, 3, 6, 74]).

Fig. 2.1 depicts the general trace-based debug architecture. At design stage, a number of internal signals are selected to be tapped. Then, during debug phase, part of the tapped signals are transferred through interconnection fabric (usually MUX tree) to on-chip trace buffer or off-chip trace port. To save DfD cost, designers can only afford to tap a small number of signals. Therefore, it is essential to select those signals that can provide a *better view* of the CUD to help designers root-cause bugs [35, 42, 78]. In addition, to save trace bandwidth requirement, trace qualification techniques (e.g., using trigger unit to control the start/stop of tracing) and trace data compression methods are often utilized [5, 24].

During post-silicon validation, electrical bugs are the most difficult to resolve because their occurrences are sensitive to certain electrical environment. In [53], the authors proposed to locate electrical errors in microprocessor by tracing the footpoint of instruction flow. However, this method cannot be applied to debugging general logic circuits. In this chapter, we consider those electrical bugs that cause the performance degradation for general logic circuits. Such bugs often occur on the critical paths in the circuit that determine its maximum operational frequency, known as speedpaths. To understand such electrical bugs and root cause them, it is essential to make the relevant signals visible during post-silicon validation.

Due to DfD cost considerations, designers can only afford to tap a subset of the relevant signals to speedpaths and trace some of them concurrently during each debug run. Consequently, the effectiveness of this debug strategy highly relies on which signals are selected to trace in the circuit. Existing trace signal selection methods (e.g., [35, 42]), however, are not applicable because they implicitly assume the timing correctness of the circuit so that the traced circuit state can be used to reason a large amount of untraced signals. At the same time, speedpath-related bugs are activated only when the corresponding speedpaths are sensitized. Therefore, if we conduct continuous tracing, the trace buffer can easily become full without any useful traced data that actually activate bugs.

The above limitations motivate the proposed trace-based debug solution for speedpath-related electrical bugs. The main contributions include

- we develop new trace signal selection algorithms to maximize the detection probability for such bugs;
- we propose novel trace qualification techniques to efficiently utilize trace bandwidth.

5.2 Observing Speedpath-Related Electrical Errors

5.2.1 Speedpath-Related Electrical Error Model

While speedpath-related electrical errors can be caused by various reasons (e.g., insufficient driving strength or excessive coupling noises), they all behave similarly as causing excessive delays on critical paths. Consider the circuit shown in Fig. 5.1 as an example. During one clock cycle in the circuit's operation, a falling transition is propagated along the path. That is, the logic value (e.g. '0' in Fig. 5.1) should be propagated through the path to the endpoint (i.e. output or flip-flop) and the value should be latched after the clock cycle. However, due to electrical bugs in the circuit, this value does not arrive at the endpoint at the end of the clock cycle, and if we observe an opposite latched value (e.g., '1' in Fig. 5.1), we can conclude that a speedpath-related electrical error occurs.

From this example, the detection of speedpath-related electrical error requires us to monitor the *propagation behavior* of the path. As for this example, to *determine* the propagation of value '0', together with the start point signal *a_in2*, we also need

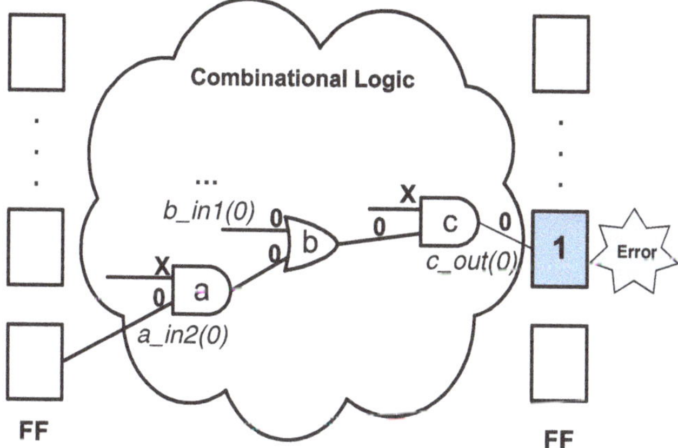

Fig. 5.1 Modeling speedpath-related electrical error: an example

to observe side-input *b_in1* of Gate *b* on the path. This is because, to determine the value propagation through each logic gate on the path, when the to-propagate value is a "controlling" value (as '0' for Gate a, c), it can definitely propagate through the gate. Otherwise, the event only happens when all other "side-inputs" (e.g., *b_in1*) are visible as "non-controlling" values (they are referred as *objective signals* hereafter). Finally, we also need to observe *c_out* for error detection. To conclude, observing logic '0', '0', and '1' on signals *a_in2*, *b_in1* and *c_out* during the CUD's normal operation implies the occurrence of electrical bugs. In other words, observing the above three signals is necessary for us to detect the electrical error that causes slow propagation of logic '0' on this path when it occurs.

5.2.2 Speedpath-Related Electrical Error Detection Quality

While several evaluation metrics for trace signal selection has been introduced in [35, 42], those works implicitly assume the timing correctness of the circuit and hence cannot be used to debug electrical errors. We therefore define a few new metrics in this work to evaluate the effectiveness of trace signals for electrical errors, as summarized in Table 5.1.

To be specific, we denote by $SV1/SV0$ the selective visibility of the signals, which obviously is '1' for traced signals and '0' for unselected ones. To note, above probabilities do not represent the probability that value '0'/'1' is *actually* observed on the node (denoted as *visibility* $V1/V0$). For example, suppose the signal remains '0' in most of time, we can barely observe value '1' on it provided it is traced. Hence, we define *visibility* with Eqs. (5.1) and (5.2), where $P1/P0$ is the probability for the signal to be logic '1/0' in functional mode. In this work, we calculate $V1/V0$ by assuming some control inputs are pre-set as '1'/'0' to insure the CUD is working in functional mode, while all other inputs are with the probability 0.5 to be value '1'/'0'.

Table 5.1 Terminologies for visibility calculation

Selected visibility ($SV1/SV0$)	For selected trace nodes, $SV1 = SV0 = 1$; otherwise $SV1 = SV0 = 0$
Functional probability ($P1/P0$)	The probability that the node is '1'/'0' in functional mode
Visibility ($V1/V0$)	The probability that the value '1'/'0' is **actually** observed on the node
Propagation visibility ($PV1/PV0$)	The probability to detect value ('1'/'0') propagation based on traced signals
Propagation occurrence probability ($POP1/POP0$)	The probability that value ('1'/'0') propagation occurs
Detection quality ($DQ1/DQ0$)	The probability that the value '1'/'0' *propagation* is detected when it occurs

$$V1 = SV1 \times P1 \tag{5.1}$$
$$V0 = SV0 \times P0 \tag{5.2}$$

With these notations, we calculate the visibilities of internal signals with forward propagation based on those observable probabilities of FFs with the following equations. Note that, these equations are based on the assumption that all inputs are independent for each logic gate.

For AND gate (a,b–input, c–output)

$$V0(c) = V0(a) + V0(b) - V0(a) \times V0(b) \tag{5.3}$$
$$V1(c) = V1(a) \times V1(b) \tag{5.4}$$

For OR gate (a,b–input, c–output)

$$V0(c) = V0(a) \times V0(b) \tag{5.5}$$
$$V1(c) = V1(a) + V1(b) - V1(a) \times V1(b) \tag{5.6}$$

For XOR gate (a,b–input, c–output)

$$V0(c) = V0(a) \times V0(b) + V1(a) \times V1(b) \tag{5.7}$$
$$V1(c) = V0(a) \times V1(b) + V1(a) \times V0(b) \tag{5.8}$$

With this information, we then introduce a metric to evaluate the error detection quality. It is important to note that the occurrence of electrical error is not predictable at design stage, while can be detected only when the value propagation behavior is monitored. We therefore resort to the propagation detection quality to evaluate the error detection effectiveness. For this purpose, the Propagation Visibility (PV) is expressed by

$$PV0/PV1 = \prod V(i) \quad i \in \{\text{start point, objective signals}\} \tag{5.9}$$

Recall the example in Fig. 5.1. The required visible signals are start point of the path and the *objective signals*. In addition, as discussed earlier the corresponding to-be-visible value of objective signal should be "non-controlling" one with its relevant gate (e.g., '0' of b_in1 with Gate b). In particular, the required visible of start point simply depends on the propagation type ('0'/'1'). With these observations, we define the Detection Quality(DQ) as the conditional probability that the value '0'/'1' *propagation* is detected under the condition that the *propagation* happens, namely,

$$DQ0 = \frac{PV0}{POP0} \tag{5.10}$$
$$DQ1 = \frac{PV1}{POP1} \tag{5.11}$$

wherein the probability that '0'/'1' *propagation* occurs (POP) is calculated by assuming all path relevant signals are independent as,

$$POP0/POP1 = \prod P(i) \quad i \in \{\text{start point, objective signals}\} \tag{5.12}$$

Clearly, the above value is 100 % when all relevant signals in the driving cone of targeted path are visible.

5.3 Trace Signal Selection

With the terminologies defined in Sect. 5.2, the trace signal selection problem studied in this section is: *Given a set of targeted speedpaths,*[1] *how to set* SV *to be '1' for a constrained number* (N_{TAP}) *of tapped signals (FFs and/or input signals), so that the circuit's total error detection quality* ($TDQ = \sum(DQ0 + DQ1)$) *for all targeted potential speedpaths is maximized.*

We solve the above problem progressively as follows. Firstly, we extract the relation between to-be-selected signals and the objective signals required to be observed for error detection (e.g., b_in1 in Fig. 5.1) to guide the selection procedure (Section 5.3.1). Based on this information, we then select a *minimum* set of signals to guarantee each targeted speedpath with *non-zero* error detection quality (Sect. 5.3.2). Finally, more signals are selected to increase error detection quality under the tapped signal quantity constraint (Sect. 5.3.3).

5.3.1 Relation Cube Extraction

Since the objective signals locate in internal combinational logic, while the to-be-traced signals are state elements surrounded by them, it is essential to extract the relationship between these two sets of signals to guide signal selection. Otherwise, we may blindly choose trace signals that have little impact on the visibility of objective signals. One straightforward thought is to conduct symbolic simulation to obtain the exact logic relationship. This method, however, is not only time-consuming, but more importantly, makes deriving the sensitization probability with a subset of the relevant signals difficult. In this work, we propose to use *relation cube* to represent the visibility for objective signals in a concise and effective manner.

[1] These speedpaths can be designated by designers or automatically extracted from the design. Considering the inaccurate delay model used in timing analysis and process variation effects, speedpath identification itself is a challenging problem, but it is beyond of the scope of this work. Interested readers may refer to [8, 13].

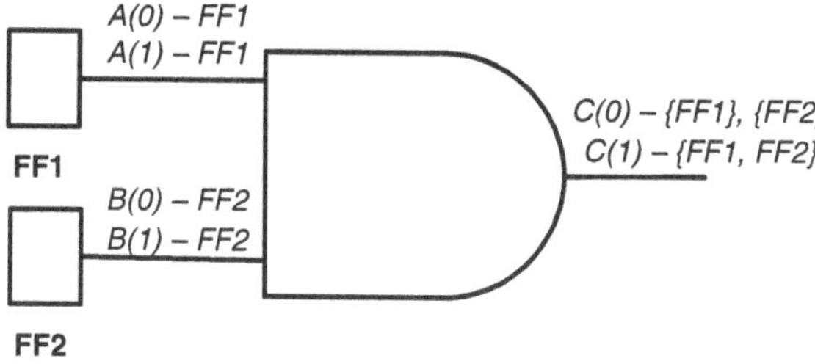

Fig. 5.2 Relation cube extraction: an example

Each relation cube denotes that when all signals in the cube are traced, the corresponding value of targeted signal can be visible. An example is shown in Fig. 5.2, where the relation cube corresponding to signal C(1) is $\{FF1, FF2\}$. We define three atomic operations (*merge*, *concatenate* and *copy*) on relation cube for its gate-level propagation. In our example, since A(1) and B(1) equals C(1), $FF1$ merges $FF2$ to be $\{FF1, FF2\}$ for observing C(1); while either A(0) or B(0) leads to C(0), $\{FF1\}$ *concatenates* $\{FF2\}$ and we have two relation cubes $\{FF1\}$, $\{FF2\}$ for observing C(0). As for *copy* operation, it is simply used when propagating the cube through an inverter or a BUFFER gate.

Based on the above, we conduct circuit-level structural analysis to extract the relation cubes for all objective signals. A pre-processing step is utilized for finding those candidate trace signals within the fan-in cones of objective signals. These signals are then initialized with two cubes corresponding to logic value '0'/'1', as shown in Fig. 5.2. We then propagate the cubes forwardly in the combinational logic. During the process, we conduct gate-level cube propagation on every newly reached logic element, until all required objective signals are processed. To note, in order to reduce the memory cost which grows exponentially with circuit size, we dynamically remove the relation cubes when the corresponding gate has fully propagated its relation cubes to all gates in its fan-out.

5.3.2 Signal Selection for Non-Zero-Probability Error Detection

After the above process, we have obtained a set of relation cubes containing candidate trace signals for observing '0'/'1' on each objective signal. With this information, this section is concerned with selecting a minimum set of signals out of all the candidates to guarantee every targeted path with **non-zero** detection quality.

From Eqs. (5.9) and (5.10), a speedpath can be monitored only when the visibilities of all its objective signals are non-zero. In other words, at least all the signals in one relation cube of each objective signal should be traced. We propose a heuristic to achieve this objective with minimum number of selected signals, and its flowchart is shown in Fig. 5.3.

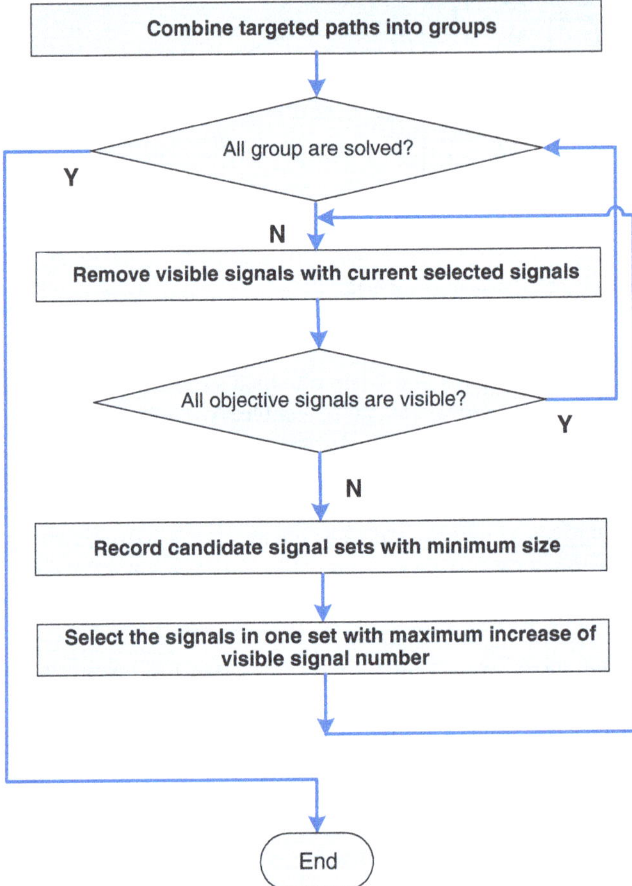

Fig. 5.3 Flow of signal selection for non-zero-probability error detection

To effectively utilize trace signals for monitoring multiple paths at the same time, those paths within the same sequential level of the circuit are put into one group to be considered together. Then for each group, the starting point and ending point of every targeted speedpath are selected, which is the basic requirement for monitoring errors on the path. Then, we gather all the objective signals and their corresponding relation cubes. In each trace signal selection iteration, we first find the set of unselected candidate signals for every relation cube, and we record the sets with minimum size from all cubes of every objective signal. Then among all these recorded sets, we choose to select the signals in one type set so that most signals will become visible. We then go back to remove newly visible objective signals and select another set of signals. The process terminates when all objective signals are visible.

With the above selection procedure, we are able to guarantee non-zero detection quality for each targeted speedpath by tracing a small number of signals. More than that, inherently the detection quality is high. This is because, we take high priority to utilize small-sized relation cubes to observe each objective signal. The probability of such event to occur tends to be high since it depends on the combination of a small number of signals.

5.3.3 *Trace Signal Selection for Error Detection Quality Enhancement*

Suppose more signals are allowed to be traced, we can use them to further improve the total error detection quality, that is, to maximize TDQ. The selection process works in a greedy manner. Because the objective signals are affected by different sets of candidate signals, we cannot evaluate the detection quality increment induced by every candidate signal. Instead, with the relation cubes extracted previously, we first determine the number of candidate signals N_{cs} that can be selected simultaneously for detection quality improvement. In other words, if the selected trace signal count is less than N_{cs}, the error detection quality is guaranteed not to increase. We obtain N_{cs} by parsing all relation cubes of the objective signals to find out the minimum missing signal number, such that if these "missing" signals are further selected, relevant relation cubes can be completed. Sequentially, we parse the cubes again to find out the corresponding missing signal sets and evaluate their impact on the error detection quality. The signals in the candidate set with the maximum TDQ increment will be selected. This procedure repeats until the total selected number reaches the predefined quantity constraint N_{TAP}.

5.4 Trace Data Qualification

Speedpaths in the circuit may not be sensitized often. Consequently, if we simply trace their relevant signals continuously, it is very likely that we end up with the data stored in the trace buffer without any useful information for error detection. We hence design a novel trace qualification module to store traced data only when slow-propagation error is found to occur on speedpaths.

The block diagram of the proposed trace qualification module is shown in Fig. 5.4. We buffer the trace signals for two cycles inside this module. When tracing for a particular speedpath, the two buffered data for its start point are firstly compared to detect whether a transition occurs on it. If not, there is no need to store the traced data. Otherwise, we rely on the slow propagation detection module to detect error, one value propagation module ('0'/'1') decided by the start point of previous cycle will assert when error is detected ("Error Assert"=1 in Fig. 5.4), and a formatter is utilized to temporarily store the traced signals and align them into trace buffer. Meanwhile,

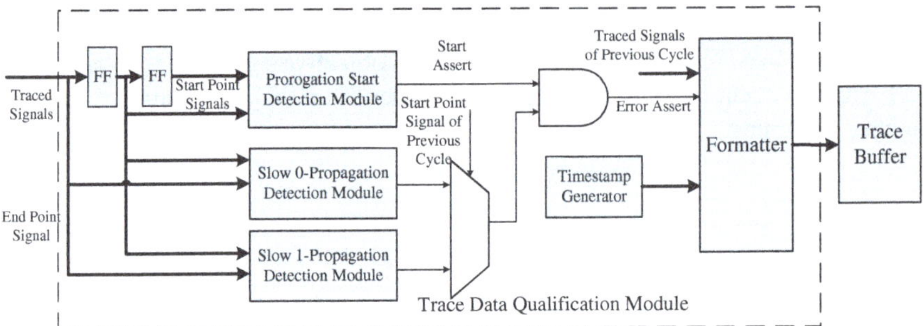

Fig. 5.4 Block diagram of trace data qualification module

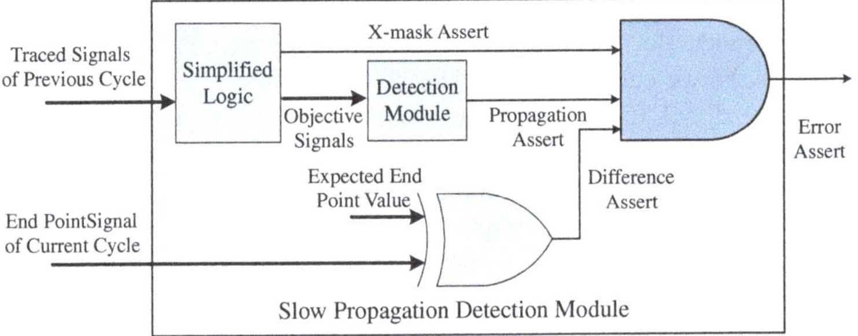

Fig. 5.5 Block diagram of slow propagation detection module

the timestamp generated from counter is also stored into the buffer for recording the error occurrence cycle.

To be specific, Fig. 5.5 describes the slow propagation detection module, which contains simplified logic and a detection block. The simplified logic can be treated as duplicating part of the CUD, while keeping all paths from traced signals to the objective ones. This is obtained by simply parsing the circuit twice, with marking propagated logic elements forwardly starting from traced signals and backwardly from objective ones. Then only relevant logic element marked by both propagations is kept in the simplified logic to achieve the above objective, as indicated in Fig. 5.6.

In addition, due to the *unknown* side-inputs that may affect the logic calculation on the kept logic (e.g. *or* gate in Fig. 5.6), we should modify the normal logic elements to facilitate "3-valued" logic calculation (e.g., X and $1 = X$). To be specific, every 1-bit wire is replaced with 2-bit one, and logic '1' is encoded as "11", '0' is encoded as "00" and unknown side-input 'X' is encoded as "01" or "10". By duplicating traced signals to 2-bit width and replacing normal logic elements with corresponding enhanced ones (designed as standard module) , we can obtain "3-valued" states on objective signals. If any one of them is 'X', it means the value is invisible from untraced relevant signals, and the "X-mask" signal asserts as '0' to denote the path is not monitored. Otherwise, when the objective signals are all required values (e.g., $b_{in1} = 0$ in Fig. 5.1) that can determine value propagation, the detection module

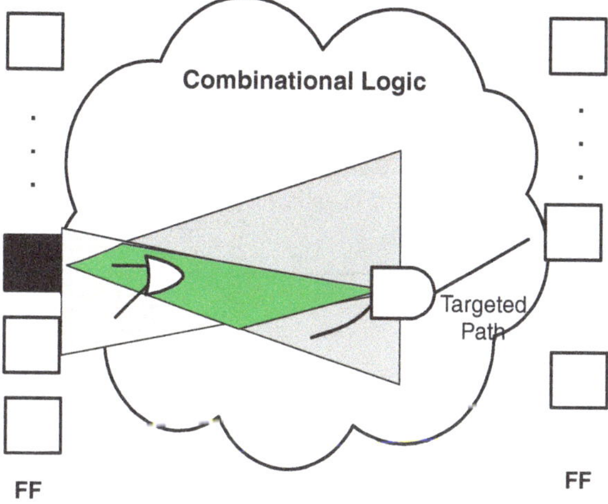

Fig. 5.6 Circuit parsing for generating simplified logic

will output '1' on "propagation assert" signal. Meanwhile if the latched value of end point in current cycle is different from the expected propagated one (obtained with logic simulation), the module asserts error signal.

Since it is not possible to trace all the tapped signals to monitor all targeted speedpaths concurrently during the debug process, we propose to implement the trace qualification module (as shown in Figs. 5.4 and 5.5) with reconfigurable logic, which is configured to detect errors on *single* targeted speedpath in each debug run. Consequently, the size of the trace qualification module is constrained by tracing a single speedpath only (the most complex one), hence reducing the associated DfD area cost. In addition, this structure can be easily extended to monitor multiple speedpaths in each debug run. Instead of simply duplicating the original module into several copies for monitoring each path, we can design the simplified logic shared by multiple targeted paths. This is feasible when several speedpaths can be grouped to share lots of logic elements in their fan-in cones. Note that, we might need to pipeline the proposed fabric to guarantee timing correctness of the monitoring circuit.

5.5 Experimental Results

We conduct experiments on several large ISCAS'89 and IWLS'05 benchmark circuits to evaluate the effectiveness of the proposed solution. We consider 50 critical paths in each circuit and we simulate them for 20,000 clock cycles. These experiments are conducted on a 2.13 *GHz* PC with 2*GB* RAM.

Table 5.2 presents the result when we select the minimum number of signals to guarantee every targeted path with *non-zero* error detection quality. Column 1 shows the name of circuit; Column 2 is the total number of state elements in the circuit;

Table 5.2 Detection quality evaluation of signal selection for non-zero visibility on 50 paths

Circuit	Total Signal #	Rel. Signal #	Sel. Signal #	UM Pro. #	Det. Pro. #	Oc. Pro. #	DQ (%)	Time (s)
s38584	1464	198	69	1	7718	9519	81.1	45.5
s38417	1664	394	157	50	630	1216	51.8	125.3
DMA	3818	1482	99	16	6827	7615	89.6	150.3
usb	2085	117	59	0	696	696	100	77.4
des	9341	132	95	46	67097	70145	95.6	668.3

Column 3 is the number of signals relevant to targeted speedpaths; Column 4 reports the number of selected signals with proposed method. We evaluate the detection quality by simulating both the original circuit and its internal behavior from the selected trace signals only, which results in a *partial* view of the circuit for each cycle. We then calculate the detection quality (Column 8) as the ratio between the detected propagation events on targeted paths (Column 6) and the total number of propagations that actually occur (Column 7). There are two different propagations on each critical path, that is, the start point can be '0' or '1'. The total number of propagation types that are completely missed to be monitored by tracing selected signals is shown in Column 5, referred to as UM Pro. # in the table. Finally, Column 9 is the CPU time.

Generally speaking, the proposed method is able to achieve satisfactory detection quality with a small portion of speedpath-related signals. For circuit usb, all propagation events on the targeted paths are captured with only 59 signals (i.e., 2.83 % of the total number of state elements in the circuit), and hence there is no need to select more signals for detection quality improvement. For circuit s38584, tracing 69 signals (34.8 % of relevant ones) guarantees all paths are visible except for one value propagation event. For circuit s38417 and des, 50 and 46 paths are unmonitored, respectively. This is because the selected signals cover a few relation cubes in objective signals, while the happened events are caused by other cubes. For circuit DMA, with only 99 out of 1482 candidate signals, we detect nearly 90 % of the propagation events, while 16 value propagations are not monitored. The selection on the largest circuit des cost 668.3s only, which demonstrates the efficiency of this procedure.

We then conduct further signal selection for detection quality improvement. The results on detection quality and unmonitored propagation quantity by incrementally selecting 10 more percent signals are plotted in Figs. 5.7 and 5.8 respectively. This procedure terminates when 95 % detection quality is reached for circuits s38584, s38417 and DMA. For circuit s38584, the detection quality approaches 100 % with only 30 % more signals (i.e., 18 signals). Meanwhile, the propagation that is not monitored in previous step becomes visible. For circuit s38417, the detection quality increases dramatically with 10 % more signals (from 51.5 to 70.8 %). Also, the number of unmonitored propagations reduces sharply during the early stage of further selection. However, the detection quality grows up to 95 % only when a great amount of signals (140 %) are traced. We attribute this phenomenon to the fact that the missed

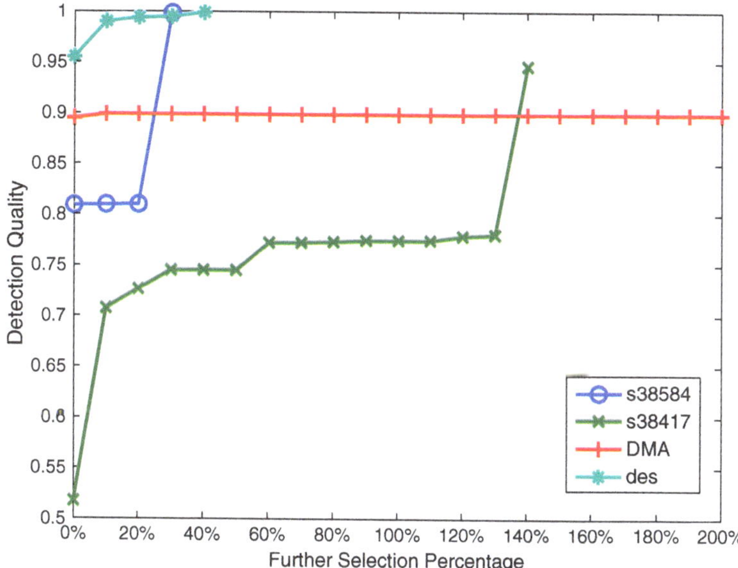

Fig. 5.7 Detection quality evaluation of improving quality selection

Table 5.3 DfD cost of data qualification module

Circuit	Trace Signal #	4-Input LUT #
s38584	87	272
s38417	367	241
DMA	108	439
usb	59	179
des	104	324

propagations of targeted paths rely on a large number of signals and it can only be detected when all of those signals are traced. The similar situation happens on circuit DMA, wherein up to 1482 signals affect the detection of targeted paths. The detection quality does not increase and the unmonitored propagation number does not decease significantly even after 200 % more signals (297 signals in total) are traced. Further selection is also conducted on circuit des, although its detection quality has exceeded 95 % beforehand. As indicated in Fig. 5.8, 42 out of 46 unmonitored propagations become visible with 10 % more signals.

Finally, we evaluate the DfD cost of the reconfigurable data qualification module. Since this module can be utilized to monitor different paths during each debug run, we consider the cost that is big enough to fit the largest detection logic for the most complex path among all the targeted ones. This module is generated automatically by parsing the circuit to obtain the simplified logic and inserting other sub-modules (e.g., formatter and timestamp generator). It is then synthesized through a commercial FPGA tool to evaluate the hardware cost. Here we choose the traced signal num-

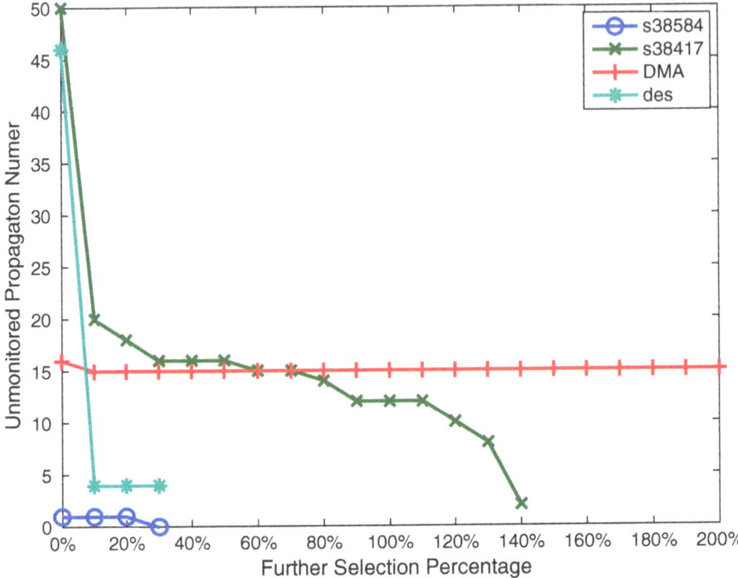

Fig. 5.8 Unmonitored propagation number evaluation of improving quality selection

ber to guarantee detection quality larger than 95 % for circuits s38584 and s38417, while keeping unmonitored propagation number to be 4 for circuit des. As reported in Table 5.3, for all cases, the maximum DfD cost is 439 4-input LUT, which is acceptable. If multiple paths required to be monitored during one debug run, the cost will not grow dramatically when these paths can share large part of logic elements in simplified logic. More importantly, the DfD cost for this module usually does not increase with the increasing number of targeted paths, because its size is determined by the path(s) with maximum requirement instead of all the targeted paths. Therefore, the relative DfD cost will be lowered when we target more paths in industrial circuits.

5.6 Conclusion

In this work, we propose a novel trace-based solution for debugging speedpath-related electrical errors, including a new trace signal selection technique that maximizes detection quality and a novel trace qualification module that improves trace buffer utilization. Experimental results on benchmark circuits show that the proposed solution are able to detect a high percentage of speedpath-related electrical errors by tracing a small number of signals with affordable DfD cost.

Chapter 6
Reusing Test Access Mechanisms

Starting from this chapter, we propose several solutions to develop low-cost interconnection fabrics (See Fig. 2.1), which are essential for ensuring debug capability. Our first work is to address one of the main difficulties in post-silicon validation. That is the limited debug access bandwidth to internal signals. Based on the observation that SoC devices often contain dedicated bus-based test access mechanisms (TAMs) that are used to transfer test data between external testers and embedded cores, in this chapter, we propose to reuse these precious TAM resources for real-time debug data transfer in post-silicon validation. This strategy significantly increases debug bandwidth with negligible routing overhead. To support different TAM architectures and debug scenarios, DfD structures are introduced at both core test wrapper level and system level. Simulation results demonstrate the effectiveness of the proposed approach at low DfD cost.

The remainder of this chapter is organized as follows. Section 6.1 reviews related work in SoC test and debug infrastructure and motivates this work. In Sect. 6.2, we present an overview of the proposed debug data transfer framework. Next, the newly-introduced DfD structures are introduced in Sect. 6.3. Section 6.4 then demonstrates how to extend the proposed method to support multi-core debug. Simulation results are presented in Sect. 6.5. Finally, Sect. 6.6 concludes this paper.

6.1 Preliminaries and Summary of Contributions

Manufacturing test and post-silicon validation are challenging problems for today's complex SoC designs. A vast body of research has been endeavored to address the above issues. We briefly survey the related work in this section and then summarize the contributions of this work.

X. Liu and Q. Xu, *Trace-Based Post-Silicon Validation for VLSI Circuits*, Lecture Notes in Electrical Engineering, DOI: 10.1007/978-3-319-00533-1_6,

6.1.1 SoC Test Architectures

Zorian et al. [79] presented a conceptual architecture for testing SoC devices, as illustrated in Fig. 6.1. The basic elements of this SoC test infrastructure include: (i) test source and sink that provide test stimuli and compare test responses (e.g., external automatic test equipment); (ii) test access mechanisms that transport test data from the source to the core under test (CUT) and from the CUT to the sink; (iii) core test wrapper that connects the core terminals to the rest of the chip and to the TAM, isolating the CUT from its environment during test.

The modular test architecture using bus-based TAMs, being flexible and scalable, are widely used in industry designs. The two most popular SoC test architectures, i.e., the Test Bus architecture [67] and the TestRail architecture [45] are depicted in Fig. 6.2. The number of TAM wires implemented on-chip varies with different SoC designs. Generally speaking, wider TAM width results in shorter testing time at a larger DfT area and routing cost. Also, large SoC designs usually implement more TAM wires on-chip to keep test cost under control. For example, 140 TAM

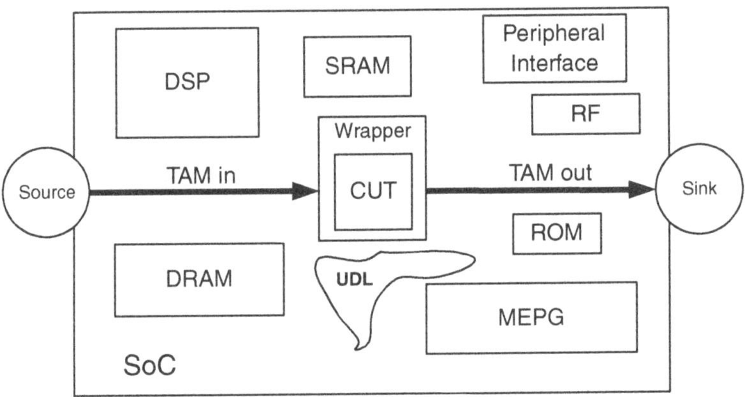

Fig. 6.1 Conceptual SoC test architecture [79]

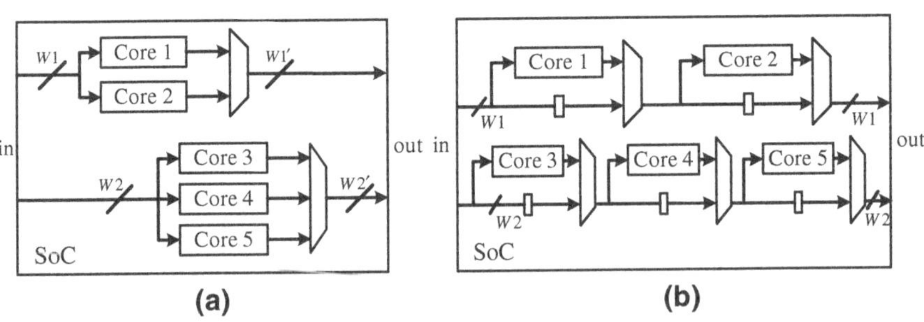

Fig. 6.2 Test Bus and TestRail architectures. **a** Test Bus architecture. **b** TestRail architecture

wires are introduced in the SoC design in [20], which are able to provide a large communication bandwidth to embedded cores.

6.1.2 SoC Post-Silicon Validation Architectures

While we expect embedded cores to work in a "plug-and-play" fashion, a few "surprises" are often inevitably discovered in first silicon and require silicon debug to identify the root causes [2]. Similar to the requirements for an efficient SoC manufacturing test strategy, an effective SoC post-silicon validation solution demands good controllability and observability of the design's internal nodes. In fact, as diagnosing an error is always much harder than detecting an error, silicon debug requires to increase these capabilities to a much higher level.

Basic postmortem debuggability can be provided by capturing snapshots of the circuit's internal sequential elements through JTAG run-control interface and scan chains [59]. This low-cost technique however is not enough for tracking tricky bugs that manifest themselves after a long period of operation. To tackle this problem, dedicated DfD structures that facilitate to trace the circuit's operations at real-time in normal functional mode (e.g., the instruction flow of an embedded processor) are implemented in most modern SoC devices. In such systems with tracing (e.g., [17, 46]), typically hardware triggers are implemented to start and stop a trace process and filter the information to be traced out.

Today's SoC devices contain an increasing number of embedded cores, and it is essential to debug the complex interactions between multiple embedded processor cores and their active peripherals. Several multi-core debug solutions have been presented in the literature and adopted by the industry. For example, in the ARM *CoreSight* debug architecture [7], each ARM core is equipped with an ETM that captures the processor's states [6]. The captured information can be stored to a trace buffer or exported immediately through an external trace port. Both require dedicated trace bus to deliver debug data from every to-be-traced core. With the ever-increasing SoC design complexity, the volume of trace data is expected to increase to identify bugs effectively, despite the use of cross-trigger and various trace qualification techniques (e.g., [5, 4, 25, 65]). Delivering such large volume of debug data to the trace buffer or trace port requires a great bandwidth. Consequently, the routing cost for dedicated trace buses is quite high.

6.1.3 Summary of Contributions

Debugging silicon is an extremely complex process, wherein the main difficulty lies in the limited visibility of the circuit's internal signals. A widely-adopted technique utilized by the industry to mitigate this problem is to reuse the IEEE Std. 1149.1 (JTAG) test access port to run, halt and step embedded cores to observe whether the values in scan chains are expected values [70]. This technique is able to effectively identify those easy-to-find bugs that leave "evidences" when the SoC halts, but fails to find trickier bugs that manifest themselves only after a long time. Therefore, an emerging trend for today's complex SoC designs is to embed more DfD structures on-chip and to monitor and trace internal signals during normal operation [3, 7]. A large amount of trace data, however, require sizeable bandwidth to transfer and most today's SoC debug architectures introduce dedicated debug buses for this duty [7, 63]. The routing of these debug buses inevitably causes significant DfD overhead to the design.

At the same time, SoC designs often contain dedicated bus-based TAMs to deliver test stimuli and responses between automatic test equipment (ATE) and embedded cores [76]. For example, 140 TAM wires are fabricated on-chip for a complex video-processing SoC device [20], making all embedded cores visible to the external ATE. These precious DfT structures, however, are usually left unused after manufacturing test. This is unfortunate because these TAM resources are able to provide a large communication bandwidth for internal signals.

Based on the above observation, in this work, we propose to reuse the existing TAMs for silicon debug data transfer [40]. This concept of reusing DfT structures for silicon debug is not new. Rather, it is similar to the strategy to reuse scan chains to "dump" data in post-silicon validation. The main difference lies in the fact that we are using these TAMs to transfer trace data at real-time. The main contributions include

- we modify the design of core test wrappers so that those to-be-observed signals in a core can be traced out at functional mode through the wrapper;
- we design DfD structures applicable for multi-core debug data transfer.

6.2 Overview of the Proposed Debug Data Transfer Framework

Before introducing the technical details of our debug data transfer framework with bus-based TAMs, let us examine the post-silicon debug flow used in our solution first, as depicted in Fig. 2.2. During the design phase, various DfD structures that support hardware triggers and core internal signals' observation are implemented. The trigger mechanisms can be simple triggers implemented with comparators and/or counters (e.g., [71]), complex cross-trigger network (e.g., [7, 65]) or even reconfigurable trigger fabrics (e.g., [2]). The to-be-traced signals embedded deeply inside a core can be brought to core boundaries by either tapping through simple multiplexer-based network (e.g., [2, 70]) or packaging in a small FIFO queue such as the ARM embedded trace macrocell [7]. During the post-silicon debug process, we first enable the aforementioned DfD structures and then put the system into normal functional mode. Once a trigger condition is hit, the traced signals are transferred along bus-based TAMs to either an internal trace buffer or an external trace port. We often need to change trigger conditions and/or trace different signals during the silicon debug process. The above operations are conducted through reconfigurations.

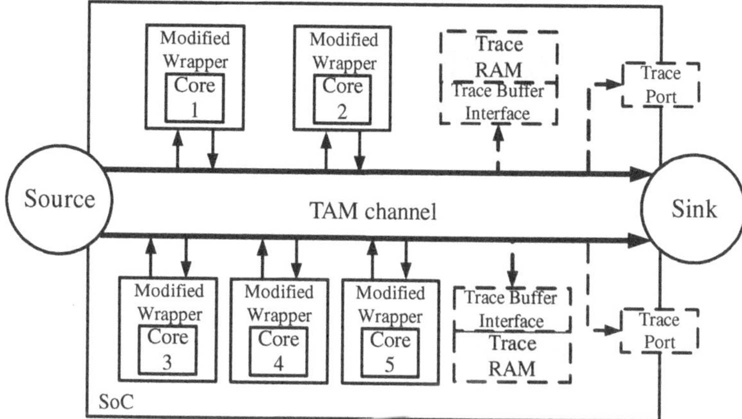

Fig. 6.3 Proposed debug data transfer framework with Bus-based TAMs

The SoC test infrastructures, however, are not designed to support real-time debug data tracing at functional mode. We need to insert some DfD structures to make this possible, as shown in Fig. 6.3. First of all, the original test wrappers need to be modified so that the traced signals can flow into TAMs in normal functional mode. Secondly, trace buffers (if any) are attached to every TAM bus to store debug data from cores under debug. As there might be multiple embedded cores on a TAM bus and they might send debug data at the same time, certain mechanisms need to be designed to avoid data corruption.

It should be also pointed out that data transfer on TAMs are controlled by a global test clock signal and its speed might be slower than that of the functional cores. Similar to transferring debug data on ARM trace buses [7], designers need to take this into account when transporting trace data on TAMs.

6.3 Proposed DfD Structures

The main objective of the proposed solution is to facilitate real-time trace data transfer through TAM channel in functional mode. To support this, the core test wrapper needs to be modified so that the *raw* debug data can be written into internal trace memory or external trace port in appropriate format in mission mode. If trace buffer is utilized, we also need to design a proper trace buffer interface to deal with TAM with various widths, as shown in this section.

6.3.1 Modified Wrapper Design

From Fig. 6.4, it can be easily observed that we have introduced a *formatter* and a *debug MUX* into the modified wrapper to conduct its duty. The debug multiplexer is utilized to select the data source of the TAM in mission mode. A new wrapper instruction "WR_DEBUG" is introduced to enable real-time debug data transfer for core under debug (CUD) at functional mode, by controlling the debug MUX. That is, when this instruction is applied, the debug data coming out of *debug output unit* will be sent onto TAM through the formatter. Otherwise, the data in the bypass register will be delivered onto TAM.

The formatter is utilized to convert the raw debug data from the CUD into a format suitable for transfer. As discussed earlier, these debug data out of the CUD's debug output unit (see Fig. 6.4) can be signals directly tapped using a multiplexer-based network (e.g., [2, 70]) or data stored in a FIFO (e.g., [7]). In our debug data transfer framework, logic '0' is put onto TAM when no CUDs are sending data to the TAM (see Fig. 6.5). Once a CUD plans to send out debug data (e.g., hardware trigger is hit), it will first send an identification tag (ID) which starts with a logic '1' so that the receiving side (e.g., trace buffer) is able to identify the start of a debug data transfer. The length of this trace process and the timestamp that the trace starts can be optionally sent onto the TAM before the actual debug data is sent out. In addition to the above, a complex formatter with a FIFO itself can temporarily store the trace

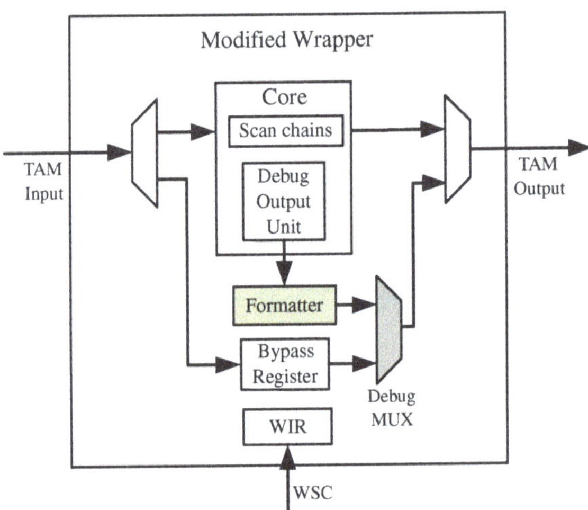

Fig. 6.4 Modified wrapper design

Fig. 6.5 Debug data transfer on TAM

data and compress them before sending them out, in order to increase the utilization rate of the TAM for debug data transfer and avoid debug data loss.

There are some differences when we configure the wrapper for Test Bus architecture and TestRail architecture, considering the case when only one core is permitted to occupy a TAM (see Sect. 6.5 for the case when multiple cores send debug data onto a TAM). In Test Bus architecture (see Fig. 6.2a), a system-level multiplexer is used to select the core that connects to TAM. Therefore, all cores can trace their debug data simultaneously and it is this multiplexer controlled by debug controller that determines which CUD can send debug data out at a particular time. For TestRail architecture (see Fig. 6.2b), however, as the TAM connects all cores in a daisy chain manner, only one core can be put in debug mode and send debug data while the other cores should be put in normal functional mode, in order to avoid data corruption because of contention.

6.3.2 *Trace Buffer Interface Design*

Figure 6.6 illustrates the structure of the trace buffer interface for the case when debug data is transferred to internal trace memories. Because the TAM width can be an arbitrary value different from the trace memory bit-width, we also need a formatter to conduct the matching. In addition, the formatter needs to detect the start of debug data transfer by identifying the non-‘0’ ID with decoder module. The data length is stored into timer and is used to count down the trace data to be written into the memory.

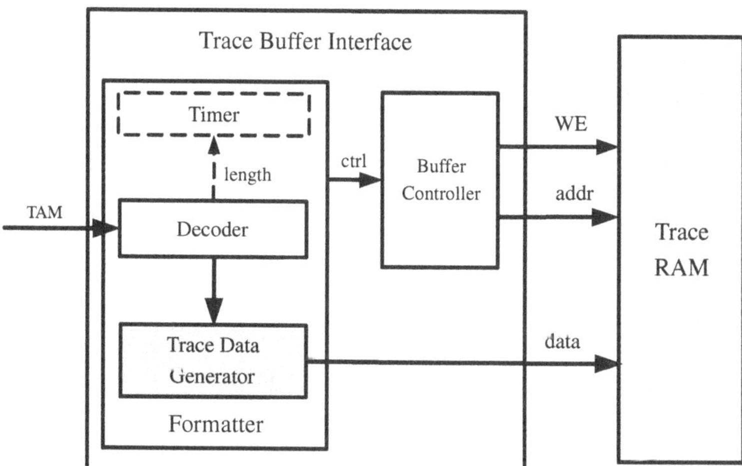

Fig. 6.6 Trace buffer interface design

In addition, we can also conduct debug data processing inside the trace buffer interface before writing them into the trace memory. For example, temporal information can be introduced and data compression can be conducted in the module of trace data generator to facilitate debug efficiency.

6.4 Sharing TAM for Multi-Core Debug Data Transfer

Since embedded cores communicate with each other during normal operation, it is often necessary to debug the complex interactions between multiple cores, especially for SoCs with many embedded processors. In the proposed debug data transfer framework, it is likely that some interacting cores connect to the same TAM and send out debug data concurrently. This is not a problem for Test Bus architecture because only one core is allowed to send data onto TAM. However, for TestRail architecture, multiple cores are daisy chained in one TAM and the debug data might get corrupted because of contention, if care is not taken. We therefore introduce a "core masking" strategy to tackle this problem. In order to support real multi-core debug that sends data onto the TAM concurrently, we propose to split the TAM into several sub-channels and these CUDs send data to different sub-channel to achieve the above objective without data corruption.

6.4.1 Core Masking for TestRail Architecture

Core masking strategy resolves the aforementioned debug data corruption problem in TestRail architecture by allowing only one core to send debug data onto the TAM at a specific time. Different from [7] that utilizes dedicated control signals to control every core, we introduce a one-bit *mask* signal generated from a *monitor* module to connect to every CUD on the same TAM. Because of the unidirectional characteristic of TestRail architecture, this monitor connects to the far end of this TAM.

The *core masking* strategy works as follows. The *monitor* unit keeps observing the activities on the TAM. Once it detects a non-'0' ID signal (i.e., a transfer request), it will assert the mask signal. For those cores that run normally and do not transfer debug data onto the TAM, the debug multiplexer in their wrappers (see Fig. 6.4) will select the data source from bypass register when detecting the assertion of the mask signal.

It should be noted that because a TAM goes through the bypass in every core on the TAM, it is possible that when one core's ID arrives at the monitor, some other CUDs on the TAM have also sent transfer requests. Debug data corruption will inevitably occur in such case. Designers can reconfigure the triggers in those CUDs to avoid this situation. A better solution however is to let those CUDs to be able to actually send data at the same time, as illustrated in the following "channel split" method.

6.4.2 Channel Split

The "channel split" strategy is to divide the TAM into several sub-channels so that every CUD that needs to send debug data concurrently can occupy a sub-channel by itself. Apparently, this capability is achieved under the condition that these CUDs' debug bandwidth requirements can be satisfied with the sub-channels, and the formatters in their corresponding wrappers need to be configured accordingly.

To apply this methodology in TestRail architecture, the debug multiplexer inside the core wrapper (see Fig. 6.4) is replaced by a multiplexor network. Figure 6.7 presents an example in which a 5-bit TAM is split into a 2-bit sub-channel and a 3-bit sub-channel for debug data transfer. Here "Test input" and "Debug input" are connected to bypass register and debug output unit in the wrapper, respectively. By doing so, only part of the TAM is occupied by this core since the other part goes through bypass register inside the wrapper. To make this happen, again, we need to introduce extra wrapper instruction. By decoding this instruction, the control signals for the multiplexor network are set to appropriate values and an example is illustrated in Table 6.1.

To apply this methodology in Test Bus architecture, in addition to the above modifications, the system-level multiplexer (see Fig. 6.2a) needs to be expanded to

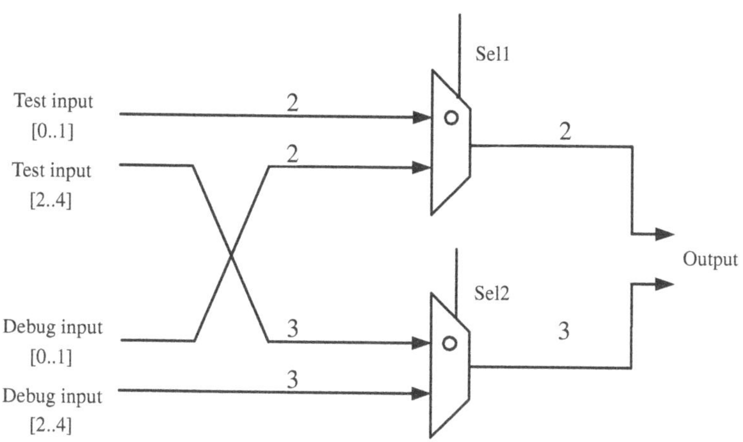

Fig. 6.7 Sharing TAM for debug data transfer with channel split

Table 6.1 Mux control for the channel split strategy

Mode	Sel1	Sel2	Output[0..1]	Output[2..4]
Other	1	1	Test[0..1]	Test[2..4]
Share (low-half)	0	1	Debug[0..1]	Test[2..4]
Share (high-half)	1	0	Test[0..1]	Debug[2..4]
Debug	0	0	Debug[0..1]	Debug[2..4]

connect the combined sub-channels to the TAM while keeping its original exclusive characteristics, controlled by the system debug controller.

While the above channel split method effectively supports concurrent multi-core debug data transfer, it might not be enough if the total debug bandwidth requirements exceed what can be provided by the TAM. In such case, we might have to add dedicated trace bus to resolve this issue.

6.5 Experimental Results

To verify the proposed debug data transfer framework, we present simulation results for two debug scenarios for a hypothetical SoC. This SoC uses a TestRail architecture and contains two 8-bit wide TAMs, as shown in Fig. 6.8. Both TAMs connect to an internal trace memory with 16-bit data bit-width. We target on debugging Core 1 and Core 2 connected on TAM_1 and Core 3 on TAM_2. The trace data bit-widths out of the debug output unit for all these three cores are 16-bit.

In our first experiment, we present simulation results for the "core masking" strategy. As shown in Fig. 6.9, after trigger request (*trigger_core1*) is activated from Core 1, the corresponding debug data transfer operation starts, which can be observed from the rising edge of write enable signal (*we_trace_RAM1*) together with the increment of address signal (*addr_trace_RAM1*). Debug data (*dataout_trace_RAM1*) are thus written into the trace memory on TAM_1. It can be also seen that the mask signal (*mask*) is asserted after detecting this event. During the debug data transfer process for Core 1, Core 2 has also detected a trigger hit (*trigger_core2* gets asserted), but because of the asserted mask signal, Core 2 could not send debug data onto TAM_1 and we can see a constant value '0x0040' that is sent from core 1. Only after this transfer process is finished, *mask* signal is de-asserted. Core 2 then occupies TAM_1

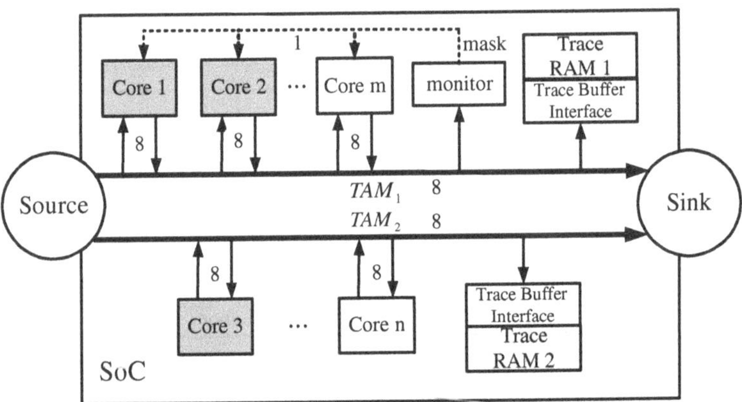

Fig. 6.8 Experimental setup

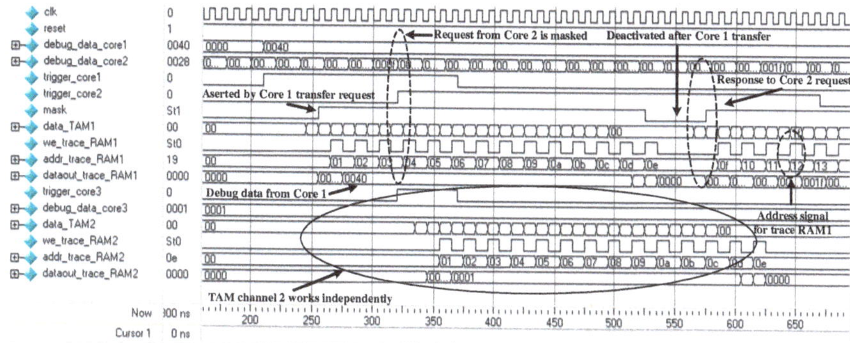

Fig. 6.9 Simulation results for core masking strategy

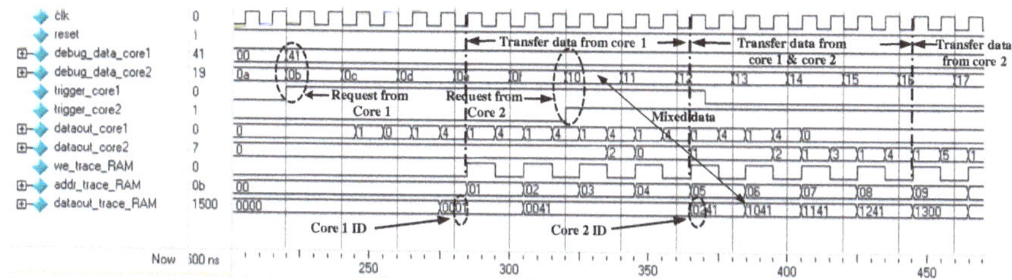

Fig. 6.10 Simulation results for channel split strategy

and starts to transfer its debug data. It can be also observed in Fig. 6.9 that, since TAM_2 is a separate channel, the debug data transfer for Core 3 works independently.

Our second experiment is to validate the *channel split* method applied on TAM_1, in which Core 1 and Core 2 share the TAM wires by utilizing the lower half and the upper half 4-bits, respectively. It should be noted to accommodate the bandwidth decrease in such case, the trace data from each debug output unit and the CUD formatter module is reduced to 8 and 4 bits, respectively. As depicted in Fig. 6.10, after trigger signal (*trigger_core1*) is asserted in Core 1, 16-bit data (*dataout_trace_RAM*) is written into trace memory, in which only half of it contains useful record information '0x41' from core 1. When the debug data transfer is also activated for Core 2 (*trigger_ core2* gets asserted), the 16-bit data now contain the debug data for both Core 1 and Core 2. As can be observed in Fig. 6.10, the 'ID' signals for Core 1 ('0x01') and Core 2 ('0x02') are first written into trace memory before the actual debug data. We can easily analyze the mixed debug data inside the trace memory according to the following rules:

$$debug\_data\_core1 = data\_RAM[7:0]$$
$$debug\_data\_core2 = data\_RAM[15:8]$$

Table 6.2 DfD area cost

Module	Area	
	Experiment 1	Experiment 2
Formatter (in wrapper)	380	289
Trace buffer interface	651	706
Debug MUX (in wrapper)	160	151
Monitor	139	

Finally, in terms of DfD cost, Table 6.2 present the silicon area of each newly-introduced DfD unit in the above two debug strategies using a commercial synthesis tool. The DfD cost of each core wrapper, trace buffer interface and monitor are all in the hundreds of 2-input NAND gates range, which is quite small. Considering the proposed technique saves the routing cost for dedicated trace bus, the total DfD cost for SoC post-silicon validation is significantly reduced.

6.6 Conclusion

In this work, we present a new silicon debug data transfer framework by reusing existing on-chip TAM resources. We propose to modify the design of core test wrappers so that core internal signals can be traced out at functional mode. We also describe two techniques that enable multiple cores to send debug data to the same TAM. Simulation results show the correct behavior of the proposed solutions.

Chapter 7
Interconnection Fabric for Flexible Tracing

In this chapter, we are going to introduce a novel low-cost interconnection fabric for improving flexibility in trace-based debug. The fabric that interconnects the large number of tapped signals to the trace buffers/ports involves non-trivial DfD overhead. Existing solutions typically use pipelined multiplexer (MUX) trees to conduct the transfer duty (e.g., [1, 34, 69]). As any signals going through the same multiplexer cannot be observed concurrently, this ad-hoc technique limits the visibility to the CUD. At the same time, since bugs often occur in unexpected scenarios, designers are typically not knowledgeable about exactly which signals should be traced together at the design stage and it is a rather cumbersome process for them to manually build the MUX network that satisfies debug needs.

In this work, we propose a novel interconnection fabric design to tackle the above problem. With the proposed method, designers are able to flexibly select any trace signal combinations (so long as they are not mutually-exclusive and do not exceed the provided trace bandwidth) in each debug process, which significantly enhances the CUD's debuggability at low DfD hardware cost. The remainder of this chapter is organized as follows. Section 7.1 reviews related prior work and motivates this work. The proposed interconnection fabric for tracing signals in post-silicon validation is detailed in Sect. 7.2. Section 7.3 then presents our experimental results on benchmark circuits. Finally, Sect. 7.4 concludes this work.

7.1 Preliminaries and Summary of Contributions

As shown in [1], for million-gate industrial designs, it is common to tap thousands of signals in the circuit and select a subset of them (say, 32 signals) to trace concurrently in each debug process.[1] These trace signals are then transferred to on-chip trace buffers and/or off-chip trace ports for later analysis.

[1] Due to the limited trace bandwidth, it is impossible to trace all the tapped signals at the same time.

X. Liu and Q. Xu, *Trace-Based Post-Silicon Validation for VLSI Circuits*, Lecture Notes in Electrical Engineering, DOI: 10.1007/978-3-319-00533-1_7,

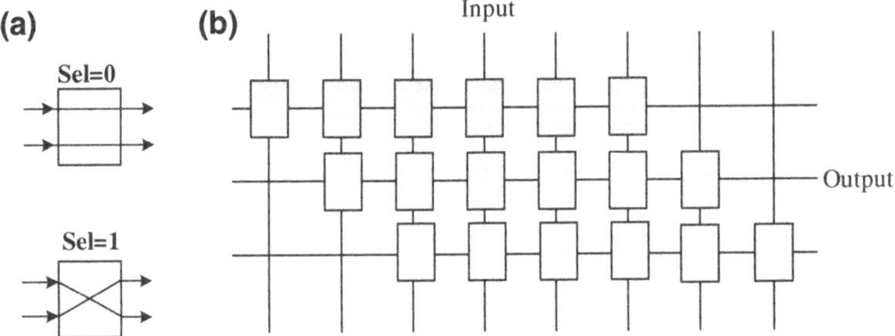

Fig. 7.1 Sparse crossbar concentrator. **a** 2×2 crossbar switch, **b** 8-to-3 sparse crossbar concentrator

The duty of the trace signal interconnection fabric shown in Fig. 7.1 is to select a subset from large amount of tapped signals and to transfer them to trace buffers/ports. Industrial designs typically use MUX trees to select a subset of the tapped signals to trace in each debug process, in which the control signals to the multiplexers can be configured through the JTAG interface (e.g., [1, 69]). To satisfying the timing constraint for the tracing logic, the MUX trees can be pipelined. In addition, when the tapped signals are coming from different clock domains, FIFO buffers and/or flip-flop chains can be used to ensure data safety [1].

To conduct root cause analysis for design bugs effectively, it is desired to have the flexibility to observe some combinations of related tapped signals (i.e. state signals and corresponding signals that may activate state transition for finite state machine) under the trace bandwidth constraint. MUX-based interconnection fabric, however, limits this flexibility and reduces the visibility to the CUD, as any signals going through the same multiplexer cannot be traced concurrently. It is hence up to the designers to manually build the MUX network based on their design knowledge, which is a rather cumbersome process for them. More importantly, since bugs often occur in unexpected scenarios, designers are typically not knowledgeable about exactly which signals should be traced together at the design stage to satisfy their debug needs.

To overcome the above limitation, designers can resort to *nonblocking concentration networks*, which have been extensively studied in the context of communication theory [54]. A n-to-m concentrator is able to transfer *any* m out of n ($m \leq n$) input signals to the output side, which naturally satisfies our interconnection fabric design needs that tries to select a subset of trace signals from a large number of tapped signals. A well-known concentrator, namely *sparse crossbar concentrator*, is a direct-connected network constructed using crossbar switches. [47] proved that the number of switches required to build such *single-stage* concentrators is at least $(n - m + 1) \times m$ and an example 8-to-3 concentrator is shown in Fig. 7.1.

Narasimha [48] proposed a multi-stage concentrator that is able to dramatically reduce the required number of crossbar switches. This *Narasimha concentrator* is developed on top of the well-known 3-stage *Clos network* [16]. Clos network is able to provide all possible permutations of the n ($n = 2^k$) inputs at the output side of the

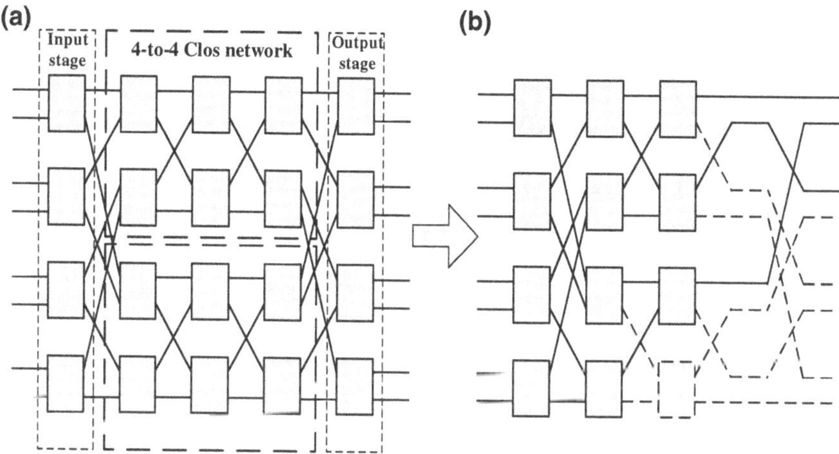

Fig. 7.2 Narasimha network. **a** 8-to-8 Clos network, **b** 8-to-3 Narasimha concentrator network

network, and it is constructed in a recursive manner. An example 8-to-8 Clos network is depicted in Fig. 7.2a. The 8 input signals are firstly assigned to four 2×2 crossbar switches, and their outputs are connected *evenly* to the upper and lower sub-networks, which are 4-to-4 Clos networks. The output stage is constructed in a reverse manner, and the whole process ends when the sub-network are all 2×2 crossbar switches. More details about the Clos network can be found in [28]. As it is not necessary to provide the "permutation" capability in a concentrator, Narasimha concentrator eliminates the output stage of the Clos network. It also removes those crossbar switches that do not drive outputs. An example 8-to-3 Narasimha concentrator is shown in Fig. 7.2b.

Recently, Quinton and Wilton [57] proposed to use the Narasimha concentrator to connect a programmable logic block to the other cores in SoC designs. While the original Narasimha concentrator requires the number of inputs to be 2^k (similar to the Clos network), it can be easily revised to take arbitrary number of inputs, as shown in Fig. 7.3. Quinton and Wilton [57] also did minor modifications to the Narasimha concentrator design to further reduce its hardware cost. To be specific, [57] proposed to replace the last crossbar unit with two direct wires at the input stage, as shown in Fig. 7.3. This replacement, however, cannot always provide the desired functionality. A counter example for a 8-to-3 network is shown in Fig. 7.3. If the last crossbar at input stage is reduced and we would like to trace inputs {0, 1, 7} at the same time, apparently, input 7 should be transferred to output 2 with the lower sub-network. Then, one of the other two inputs cannot find a path to the output. Consequently, the combination of inputs {0, 1, 7} cannot be traced together, which violates the functionality for non-blocking concentrator design.

Apparently, using concentrators to construct trace signal interconnection fabric provides better visibility to the CUD when compared to MUX trees, but at the cost of more DfD area. In practice, some tapped signals are not highly correlated with each

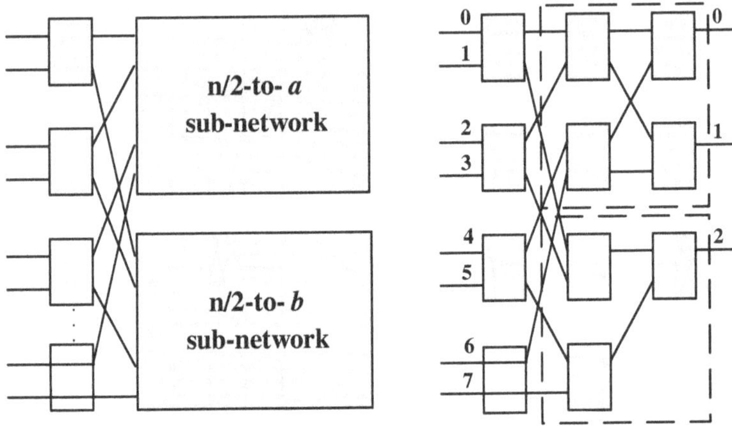

Fig. 7.3 Revised Narasimha concentrator in [57]

other, and hence it is not necessary to observe them concurrently. This observation motivates us to combine MUX trees and concentrators to have a flexible yet low-cost tracing network. In addition, even for the revised Narasimha concentrator, it still contains lots of redundant elements. Take a special case for 8-to-8 network as an example, the simplified concentrator requires 9 crossbar switches to construct it, but these switches actually can all be eliminated. This motivates us to propose a new concentrator design with much lower hardware cost. We detail the proposed interconnection fabric in the following section. The main contributions of the fabric include

- we use *MUX network* that connects those mutually-exclusive tapped signals, which can be designated by designers and/or extracted automatically based on structural analysis;
- we develop simplification method on *non-blocking concentration network* that is able to transfer any m signals out of n inputs ($m \leq n$) to the trace buffers/ports.

7.2 Proposed Interconnection Fabric Design

The problem investigated in this work can be formulated as:

Given N_{tp} tapped signals in the CUD, those highly-correlated signals should be able to be traced concurrently, while others not. We are to design an interconnection fabric to transfer a subset of N_{bf} (typically, $N_{bf} \ll N_{tp}$) tapped signals to the trace buffers/ports at runtime, which satisfies the above requirement at the minimum hardware cost.

To tackle the above problem, we propose to design the interconnection fabric as shown in Fig. 7.4, which contains two main parts: (i) a *multiplexer network* that connects those mutually-exclusive tapped signals, which can be designated by designers

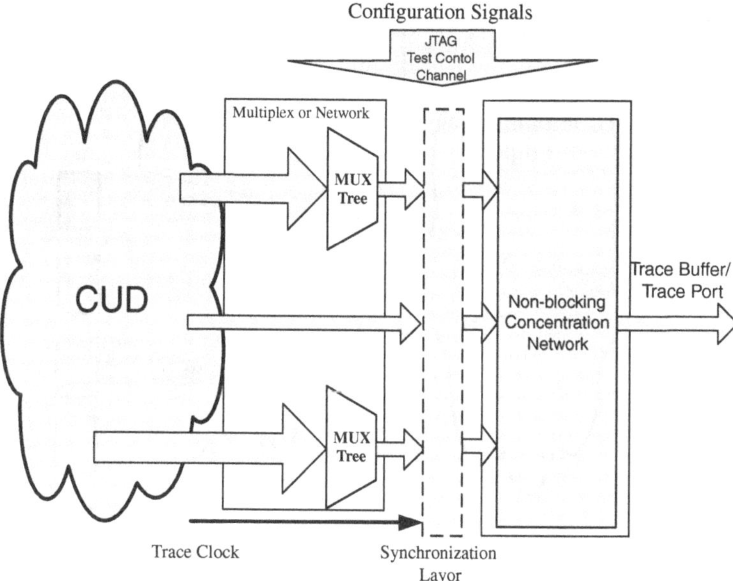

Fig. 7.4 Proposed interconnection fabric

and/or extracted automatically based on structural analysis. This stage outputs N_{pc} potentially-correlated signals. (ii) a *non-blocking concentration network* that is able to transfer any N_{bf} signals out of N_{pc} inputs to the trace buffers and/or trace ports. One of the main objectives of our design is to minimize N_{pc}. The reason behind this is that, accessing the same signals with MUX tree always results in smaller DfD cost when compared to using nonblocking concentrator, since the latter one requires more resources to achieve the "any combination" objective.

7.2.1 Multiplexer Network for Mutually-Exclusive Signals

Firstly, we need to determine which tapped signals are highly-correlated and hence may need to be traced together in post-silicon validation. As discussed earlier, it is a rather cumbersome process for the designers to manually conduct this duty. We introduce a simple yet effective method to facilitate this process through circuit structural analysis. For a tapped signal, its *related logic elements* are those that are either in the logic cone that drives this signal starting from inputs or in the logic cone that this signal drives until outputs. For two tapped signals, if there is no overlap between their respective related logic elements, they are not correlated at all. The above constraint is quite stringent and it does not reflect how high the correlation is among tapped signals. We therefore first levelize the sequential elements of the circuit. Obviously, the closer a logic element is to the tapped signal, the more related

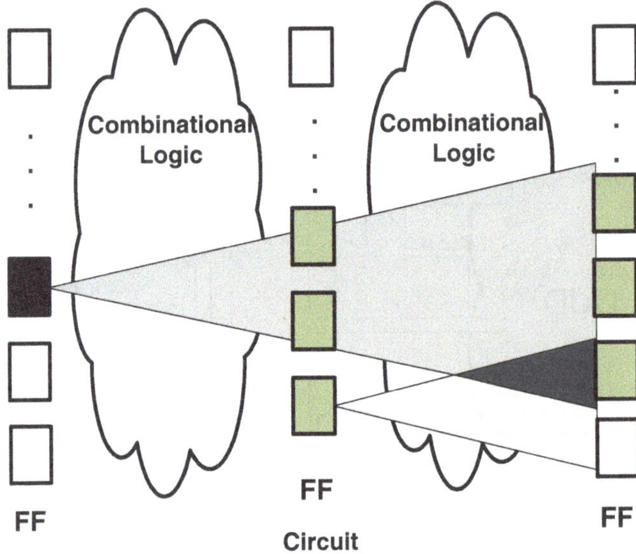

Fig. 7.5 Forward propagation for correlation extraction

it is with the signal. Then, instead of checking the overlapping of all the related logic elements of two tapped signals, we only check whether there is any overlap within the neighboring N_l levels, in which N_l is a user-defined value. If there is, we call these tapped signals *highly-correlated* and should be able to be observed together. Otherwise, they are *mutually-exclusive*.

According to the above, we build an *uncorrelation graph* among tapped signals. In this graph, each vertex represents a tapped signal and we add an edge between two vertices if they are not highly correlated. The graph is initialized as a complete graph. The edges are then gradually removed from the graph by conducting forward propagation analysis for the circuit from the first sequential logic level, as shown in Fig. 7.5. Note, there is no need to conduct backward analysis for a tapped signal, since the correlations are already obtained by those tapped signals in previous logic level through forward propagation, if any.

Our MUX network is composed of a number of MUX trees (see Fig. 7.4), which are used to connect mutually-exclusive signals and only one of the signals is required to be observed from each MUX tree at a time. In our uncorrelation graph, each MUX tree corresponds to a *clique*. To minimize the number of outputs N_{pc} from the MUX network, it is equivalent to use the minimum number of cliques to cover all the vertices in the uncorrelation graph. This "minimum clique cover problem" is a well-known NP-hard problem in graph theory, and we can resort to a classical greedy heuristic to solve it [21]. As shown in Fig. 7.6, at least two cliques are required to cover all 6 nodes, and two MUX trees are introduced accordingly to merge these signals and $N_{pc} = 2$.

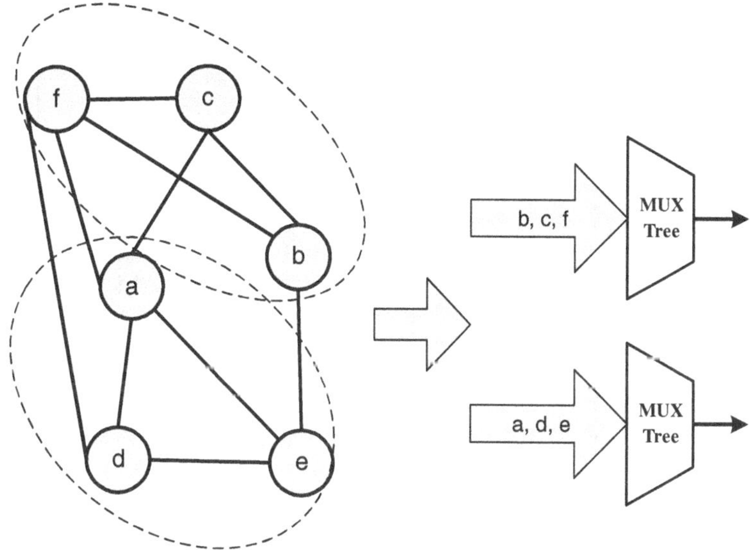

Fig. 7.6 An example of uncorrelation graph

The above discussions are applicable when all tapped signals are from the same clock domain. If, however, we have tapped signals from different clock domains to transfer to the same trace buffer and/or trace port, we need to build the uncorrelation graph separately for signals from each domain and we need to add a synchronization layer before these signals going into the non-blocking concentrator (see Fig. 7.4). At the same time, in order not to miss any traced data, it is important to trace signals using the fastest clock among all domains.

7.2.2 *Non-Blocking Concentration Network for Concurrently-Accessible Signals*

Starting from the revised Narasimha concentrator that is able to take arbitrary number of inputs as shown in [57] (without using their simplification method as it violates our desired concentration capability, see Sect. 7.1), we propose several simplification rules to reduce the DfD cost of the concentration network, detailed in the following.

- **Rule 1: Replacement of crossbars that provide redundant path**

As mentioned before, the structure presented in existing work provides redundant paths where signals will never go through. We first introduce a theorem for the output port assignment of the revised Narasimha network, which is missed in [57]. This theorem guarantees that any m out of n input signals are accessible by this concentrator.

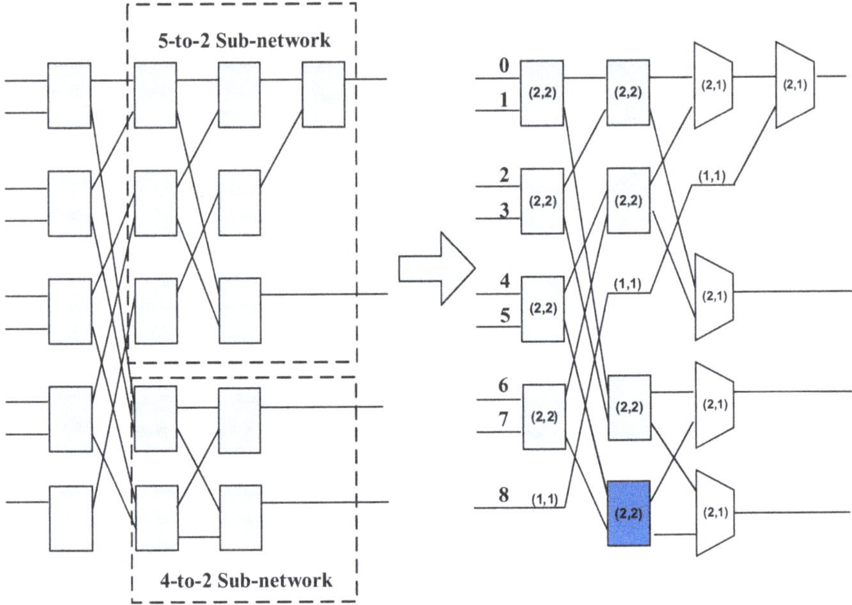

Fig. 7.7 9-to-4 network example of Rule 1

Theorem 1 For any n-to-m Narasimha-based concentrator, if the m output nodes are evenly distributed into the top half and the bottom half of the concentrator, it is able to provide n to m accessibility.

Proof Any two input signals of the crossbar can be connected to the top half and the bottom half of the concentrator, respectively. Therefore, any m out of n input signals can evenly flow into two sub-fabrics with no larger than one difference (i.e., $\{k+1, k\}$ or $\{k, k\}, k = \lfloor n/2 \rfloor$). Recursively, every sub-fabric also has *non-blocking* feature. As a result, if the output nodes are assigned accordingly, the accessibility is guaranteed. □

To propagate the simplification effect, we firstly assign the input and output signal effect on each switch. As depicted in Fig. 7.7, every switch has a (Input Effect, Output Effect) initialized as $(0, 0)$. Then Input/Output Effect of switches in input/output stage is updated as the number of assigned input/output signal. After that, the Input/Output Effect are propagated forwardly/backwardly. The forward propagation is as follows. Consider one switch at middle stage, if its Input Effect is 2, two back-end connected switches will add 1 for their own Input Effect, since both paths are required to transfer signals. Similarly, when its Input Effect is 1, then only one connected switch will add 1 for its Input Effect (here we choose the upper one). The backward propagation is conducted in the same way but with the opposite direction.

After both propagations, the redundant units in the original concentrator can be identified and simplified as shown in Fig. 7.7. If both Input Effect and Output Effect are 2 for a switch, it remains to be a crossbar unit. If they are 2 and 1 respectively, the switch is simplified to be a MUX. Otherwise, it is replaced by a wire. With the above simplification rule, for a 9-to-4 concentration network, its cost will be reduced from 16 crossbars to 8 crossbars and 5 MUXes.

- **Rule 2: Reduction of crossbar at input stage**

As discussed earlier, [57] proposed to replace the last crossbar unit at the input stage with two direct wires, but it cannot always hold the desired property for concentration network. The following theorem presents the condition for such replacement.

Theorem 2 For any n-to-m network, a crossbar switch at the input stage switch can be replaced by two direct wires, provided both n and m are even.

Proof We start from the case that n is even. By Theorem 1, at the input stage, m out of n signals can be evenly distributed into two sub-networks (i.e., $n/2$-to-a and $n/2$-to-b, where $|a - b| \leq 1$). We consider the to-be-reduced unit as the one whose two inputs are connected directly to two sub-networks, given m is even. If neither signals from two sub-networks are traced, the replacement does not affect the original transfer capability. If one of them is traced, other units at this stage is able to evenly distribute $m - 1$ signals. If both are traced, since they are separated into two sub-networks, the functionality is reserved.

However, if m is odd (say, $a = (m - 1)/2$ and $b = (m + 1)/2$), suppose the signal connected to the sub-block with a outputs is traced, the other $m - 1$ signals should be connected to the sub-networks with $a - 1$ and b outputs respectively. Since $b - (a - 1) = 2$, this requirement cannot be satisfied.

Similarly, when n is odd, if we trace the last signal (e.g., Input 8 in Fig. 7.7) and an input that also connects to the top half sub-network (Input 6), the replacement is not applicable. □

We apply this rule to further simplify the fabric at the input stages. For the 9-to-4 network depicted in Fig. 7.7, only one switch can be reduced in a 4-to-2 sub-network. To note, this rule does not work for the special case that the input stage contains only one switch.

- **Rule 3: Replacement of crossbars that provide unnecessary permutation**

Even after applying previous simplification rules, the structure still unnecessarily cost large amount of crossbar units for n-to-m concentrator, when m is large. Let us use a 8-to-7 network shown in Fig. 7.8 to demonstrate our simplification rule. Starting from the last stage, it is obvious that the switches here can only change the order of two signals to provide unnecessary "permutation". Consequently, these crossbars can be replaced with two wires as depicted in Fig. 7.8b. Next, from these direct wires, the frontend crossbar switch may be directly connected to output ports with two wires too. Based on the same principle, it can be further replaced and we continue to process the frontend crossbar switches until the input side is reached. The result for this example is shown in Fig. 7.8c and we are able to remove four crossbar switches.

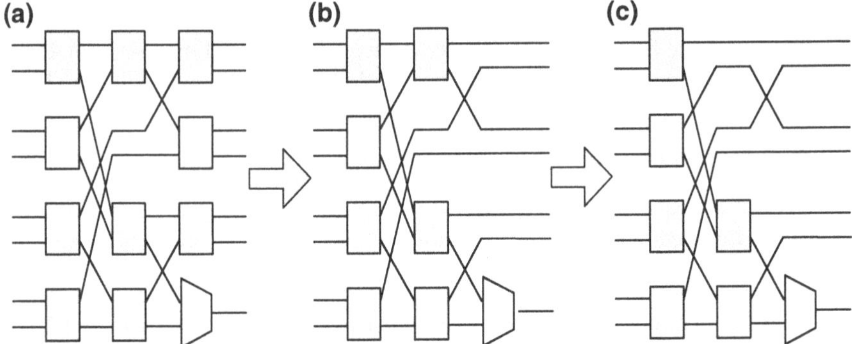

Fig. 7.8 8-to-7 network example of Rule 3. **a** Structure after Rule 1 to 2, **b** Structure after replacement in the last stage, **c** Structure after Rule 3

In particular, for the special case with n-to-n structure, after applying this rule, no crossbar units is required.

7.3 Experimental Results

We conduct experiments on three large ISCAS'89 benchmark circuits to evaluate the DfD cost of the proposed interconnection fabric and compare against existing structures. The correlation constraints are extracted by using the methods presented in Sect. 7.2.1 with $N_l = 3$. We randomly select 50, 100, 150, and 200 tapped signals and automatically generate interconnection fabric to connect those signals to trace buffers with various widths. The fabrics are inserted into the benchmark circuits and the DfD cost is obtained by using a commercial synthesis tool.

Table 7.1 shows the DfD cost introduced by the proposed approach and three existing methods. Column 1 presents the number of tapped signals; Column 2 is the proportion of the edges among all possible connections in the uncorrelation graph; Column 3 is the number of signals output from our multiplexer network; Column 4 is the trace buffer width; Column 5–8 are the DfD area cost in terms of 2-input NAND equivalent gates; and Column 9 shows the DfD area reduction resulted from the proposed method, when compared with that in [57] (again, without applying their simplification method).[2] In this table, *MUX* refers to the MUX tree that does not consider signal correlations. It indicates the lower bound for the DfD area cost to connect N_{tp} signals to N_{bf} outputs. *Sparse* corresponds to the sparse crossbar concentrator design in [48].

As expected, MUX tree and sparse crossbar concentrator costs the least and the most DfD area, respectively. With the growth of the tapped signal count, generally

[2] area overhead reduction $\Delta = (1 - \frac{\text{Area of proposed fabric}}{\text{Area of [57]}}) \times 100\,\%$

Table 7.1 Experimental results for DfD area cost

Input #	Edge (%)	N_{pc}	Buffer width	DfD Cost (2-input NAND gates)				
				MUX	Sparse	Quinton and Wilton [57]	Prop.	Δ (%)
				s35932				
			8	384	7224	2571	651	74.7
50	94.1	13	16	339	11760	3177	339	89.3
			32	339	12768	3177	339	89.3
			8	828	15624	5280	1101	79.2
100	95.2	13	16	789	28560	6192	789	87.3
			32	789	46484	7380	789	89.3
			8	1275	24024	8157	2055	74.8
150	93.2	28	16	1215	45476	9909	2067	79.1
			32	1110	80084	11223	1110	90.1
			8	1728	32424	10740	2589	75.9
200	95.0	32	16	1662	62276	12660	2685	78.8
			32	1524	113692	14484	1524	89.5
				s38584				
			8	1533	8399	3746	3210	14.3
50	32	40	16	1461	12935	4334	3474	19.8
			32	1352	13943	4352	3330	23.5
			8	4444	19234	8871	6456	27.2
100	44	68	16	4372	32170	9783	6888	29.6
			32	4236	49980	10971	7080	35.5
			8	4301	27106	11215	8213	26.8
150	39	104	16	4245	48429	12967	9149	29.4
			32	4096	83014	14281	9677	32.2
			8	6466	37197	15514	10840	30.1
200	40	143	16	6402	66907	17433	12091	30.6
			32	6263	118335	19257	13249	31.2
				s38417				
			8	528	7374	2721	1233	54.7
50	69.8	27	16	462	11910	3309	1266	61.7
			32	363	12918	3327	363	89.1
			8	1280	16047	5239	3036	42.1
100	64.6	60	16	1208	28983	6119	3384	44.7
			32	1026	46794	7307	3492	52.2
			8	1923	24649	8782	4049	53.9
150	67.6	74	16	1851	46039	10534	4607	56.3
			32	1707	80647	11848	4883	58.8
			8	2314	33025	11341	5622	50.4
200	64.6	109	16	2242	62784	13261	6495	51.0
			32	2098	114215	15085	7233	52.1

speaking, more DfD area is required for the interconnection fabrics (e.g., s38417 case shown in Fig. 7.9). When the number of tapped signal is fixed, the cost of MUX tree slightly decreases with the increase of buffer width, because less signals are required to be concentrated. By contrast, we observe proportional increase of

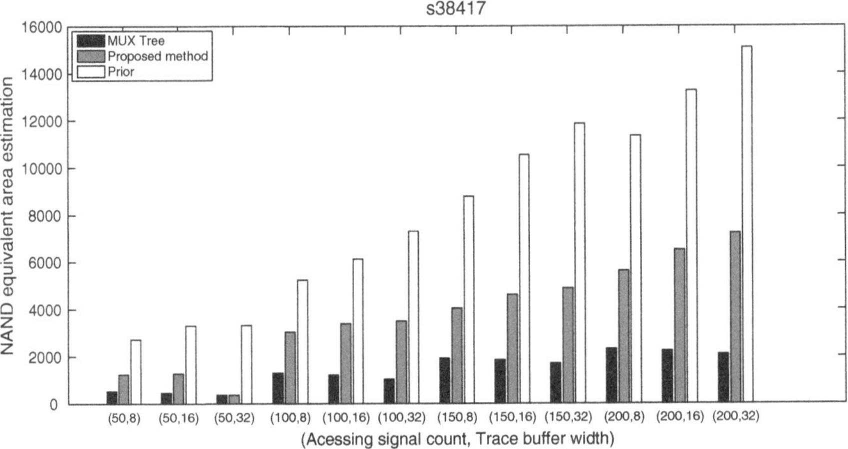

Fig. 7.9 Experimental results for DfD area cost of s38417

the sparse concentrator DfD area with various buffer widths in almost all cases. Similarly, the DfD cost using [57] grows steadily. For the proposed interconnection fabric, however, its area cost depends on not only the signal count and buffer width but also the correlation constraints. For the case that signals have low correlations (e.g., circuit s35932), the number of signals feeding into the non-blocking concentrator is usually quite small, as they have been processed in the MUX network stage. In this case, the cost of the proposed interconnection fabric can be significantly reduced, without sacrificing the debug flexibility. Taking s35932 as an example, as there are very few correlations among signals, the DfD cost of our interconnection fabric is the same as or slightly higher than that of MUX tree. In average, the DfD cost is reduced up to 83% when compared with [57]. For the case that traced signals are highly related (e.g., s38584), the DfD cost for the proposed design reduces for roughly 27 % by applying the simplification rules on non-blocking concentrators. One thing to be noted is that DfD area cost reported in this table seems to be relatively high when compared to the original circuit size, we attribute this to the large percentage of tapped signals (in practical designs, the percentage is smaller [1]).

To investigate the impact of the number of propagation levels N_l during correlation extraction, we conduct a case study for s35932 when tapping 200 signals and transferring to 16-bit trace buffer. When we set N_l to be 6, N_{pc} increases to 68 and the DfD area cost grows to 4197 2-input NAND gates. If $N_l = 12$, the DfD area cost further grows to 7809 gates. From the above, we can see that building correlation constraints among signals has a significant impact on the cost of the interconnection fabric. In practice, designers should carefully select N_l based on their design knowledge.

We also study the effectiveness of each simplification rule on non-blocking concentrator. In this experiment, we trace 100 signals in s38584 and build the interconnection fabric using the concentrator only. As depicted in Fig. 7.10, Rule 1 is the most

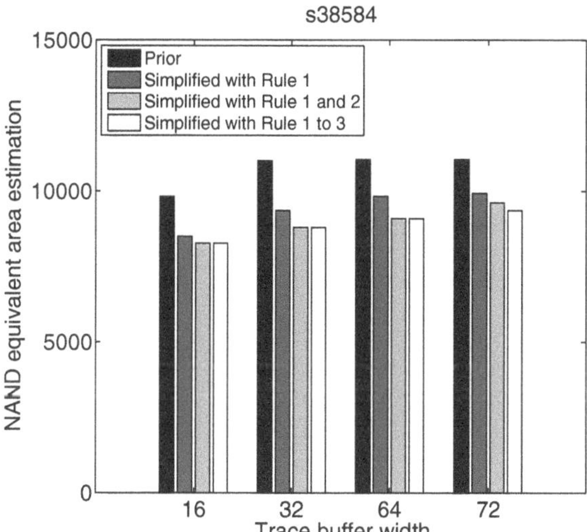

Fig. 7.10 Experimental results for DfD area cost of s38584 for simplification rule evaluation

effective one when the fabric contains 32 outputs, because the original fabric contains most redundant paths at this moment. Rule 2 saves most area for 64-bit buffer. It is more effective than 16-bit and 32-bit cases because more crossbar switches at input stages that have not been replaced by MUXes can be reduced by this rule. It is more effective than 72-bit because the rule does not work for the sub-network containing an odd number of outputs. The impact of Rule 3 is observable only when the output number is 72, since only under such circumstances the switch at the last stage of the fabric is connected with two outputs.

Finally, to address the timing issue, we can pipeline the proposed interconnection fabric by inserting flip-flops into it. Consider the case that tracing 200 signals with 8-bit buffer in s38417. The original proposed fabric introduces a few critical paths. After pipelining, these critical paths can be eliminated, with additional 3199 gates.

7.4 Conclusion

In this chapter, we propose a novel trace signal interconnection fabric design that contains a multiplexer network that connects those mutually-exclusive tapped signals and a novel simplified non-blocking concentrator. Experimental results on benchmark circuits show that the proposed solution is able to significantly reduce DfD area cost while satisfying designers' debug flexibility requirement.

Chapter 8
Interconnection Fabric for Systematic Tracing

Chapter 7 introduces a trace interconnection fabric design to improve flexibility in trace-based debug. In this chapter, we propose a trace interconnection fabric design that is able to support systematically localizing erroneous signals with just a few debug runs, at the cost of little extra DfD hardware. Experimental results on benchmark circuits demonstrate that the proposed trace interconnection fabric facilitates more efficient silicon debug than existing fabrics.

The remainder of this chapter is organized as follows. Section 8.1 reviews related works and presents the motivation of our work. In Sect. 8.2, we present the proposed flexible interconnection fabric design. The systematic error localization methodology is then detailed in Sect. 8.3. Experimental results on benchmark circuits are shown in Sect. 8.4. Finally, Sect. 8.5 concludes this work.

8.1 Preliminaries and Summary of Contributions

With the ever-increasing design complexity and the ever-shrinking market window for today's IC products, it is increasingly difficult to guarantee the correctness of the design solely through pre-silicon verification, requiring post-silicon validation to catch bugs left in the design [23]. During silicon debug (see Fig. 8.1), designers try to feed certain test input vectors that can activate the bug and observe their error effects to identify them. Since the circuit under debug is a piece of silicon that has already been fabricated, the main challenge is that there is only limited visibility of its internal signals. Consequently, one of the key issues in silicon debug is how to efficiently capture the error evidences, so that designers can quickly localize the error within a small region of the CUD, and then apply various diagnosis methods (e.g., [15]) to root-cause the bug and fix it.

One widely-used silicon debug technique is to reuse the CUD's existing test structure (e.g., scan chains) to run/stop its operation and observe whether the values in the circuit's storage elements are the expected values [31, 59]. Even though effective

X. Liu and Q. Xu, *Trace-Based Post-Silicon Validation for VLSI Circuits*, Lecture Notes in Electrical Engineering, DOI: 10.1007/978-3-319-00533-1_8,

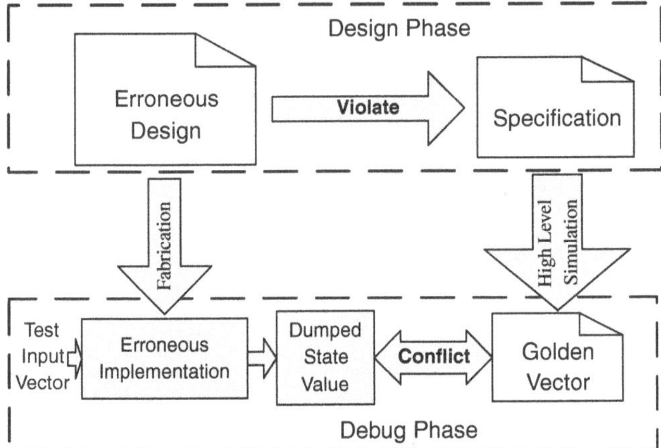

Fig. 8.1 Silicon debug: an overview

for identifying those easy-to-find bugs that leave "evidences" when the circuit halts, this low-cost technique provides little help for tracking those tricky bugs that takes a long period of operation to manifest themselves. Moreover, the behavior of many bugs is not repeatable, making diagnosis with this run/stop debug methodology even more difficult.

Tricky errors may take a long period to be activated in "corner cases" or certain electrical environment [14]. After activation, however, the error will leave its evidences in the circuit. As demonstrated in Fig. 4.3, the evidences will further propagate forwardly and leave their tracks on some FFs (e.g., the FFs in bold in Fig. 4.3) [44]. Based on the above observation, if we are able to observe the error evidences by properly tracing relevant FFs for some time, the region with error is greatly zoomed in, and hence the bug root-cause effort is significantly reduced.

One important issue in trace-based silicon debug is how to select tapped signals, which determines the effectiveness of the debug solutions, i.e., whether we are able to effectively observe bugs' erroneous effects in the circuit. Many solutions have been proposed in the literature to tackle this problem [35, 42, 56, 62, 78].

On the other hand, due to the limited trace bandwidth, we are not able to trace all the tapped signals concurrently in each debug run. The trace interconnection fabric design and the corresponding signal tracing methodology therefore determines how efficient we can observe the bugs' erroneous effects (if any), e.g., the number of debug runs to locate them.

One widely-used trace interconnection fabric is the MUX-based fabric, wherein we trace a subset of the tapped signals each time until we find out the erroneous effects. An example MUX-based fabric is shown in Fig. 8.2, wherein we need four debug runs to observe the erroneous effects in $a4$.

To reduce the number of debug runs during silicon debug, [77] proposed an XOR-based trace interconnection fabric (as depicted in Fig. 8.3), wherein all tapped signals

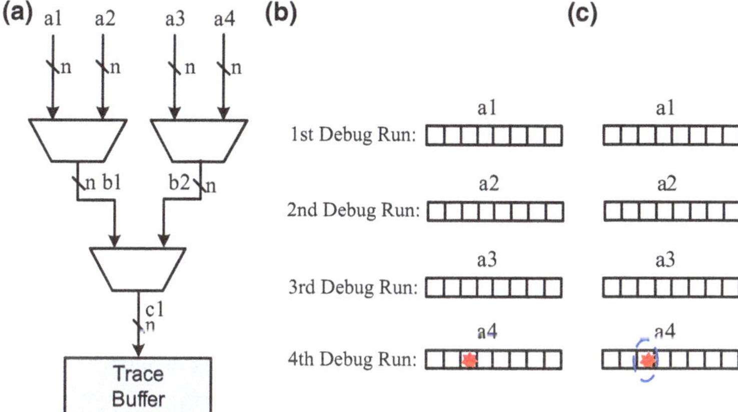

Fig. 8.2 MUX-based trace interconnection fabric design. **a** MUX Network, **b** Trace Data, **c** Localized Error Evidence

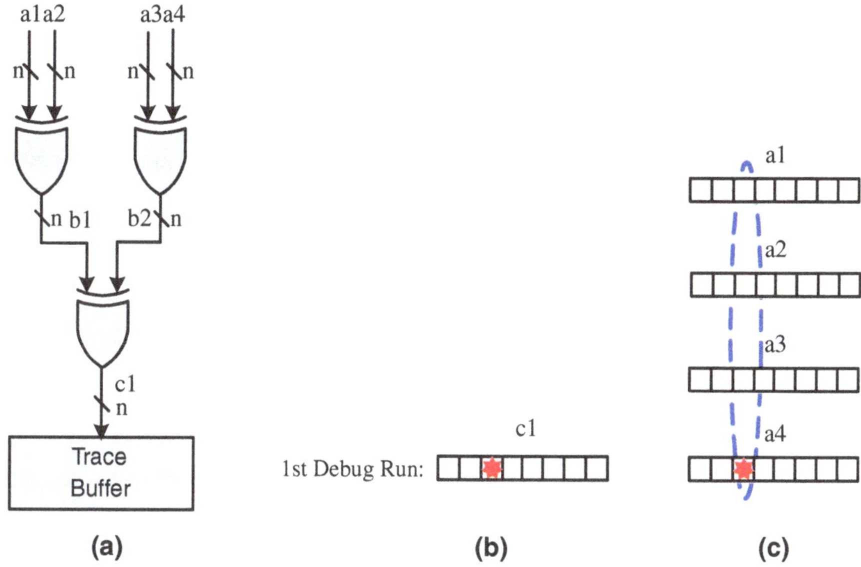

Fig. 8.3 XOR-based trace interconnection fabric design. **a** MUX Network, **b** Trace Data, **c** Localized Error Evidence

are compacted into signatures[1] and then transfer to trace buffers/ports. For the same example, with XOR-based trace interconnection fabric, we can observe erroneous evidence in the traced signature in the first debug run, but we have four suspicious tapped signals and further diagnosis is needed to identify the actual one (see Fig. 8.3). Meanwhile, the effectiveness of the XOR-based fabric relies on the existence of *clean*

[1] The likelihood for aliasing is quite small, and for the sake of simplicity, it is ignored in this chapter.

"golden vector" to generate reference signatures for comparison. However, this is usually not the case during silicon debug, rendering XOR-based fabric less effective. This is because: (i) it is often too time-consuming to run gate-level simulation for failed silicon test,[2] designers often resort to high-level simulator to generate "golden vectors" and many unknown bits (X-bits) are obtained when they are mapped onto gate-level vectors; (ii) asynchronous clock domains and uninitialized state elements also result in many X-bits in the vectors.

As indicated above, both MUX-based and XOR-based trace interconnection fabric have their own advantages and disadvantages. *A relevant question is whether we can design a hybrid fabric and its corresponding tracing methodology that have the benefits of both solutions while overcoming their limitations?* This has motivated the proposed technique in this chapter. The main contributions include

- we present a hybrid trace interconnection fabric that is able to tolerate unknown bits in "golden vectors", at the cost of little extra DfD overhead;
- we introduce a systematic signal tracing procedure to automatically localize erroneous signals with just a few debug runs.

8.2 Proposed Trace Interconnection Fabric

Our first thinking is to design a straightforward hybrid trace interconnection fabric as depicted in Fig. 8.4. In addition to the XOR network used to trace compacted signatures, a cross-connected MUX network is used to trace tapped signals directly.

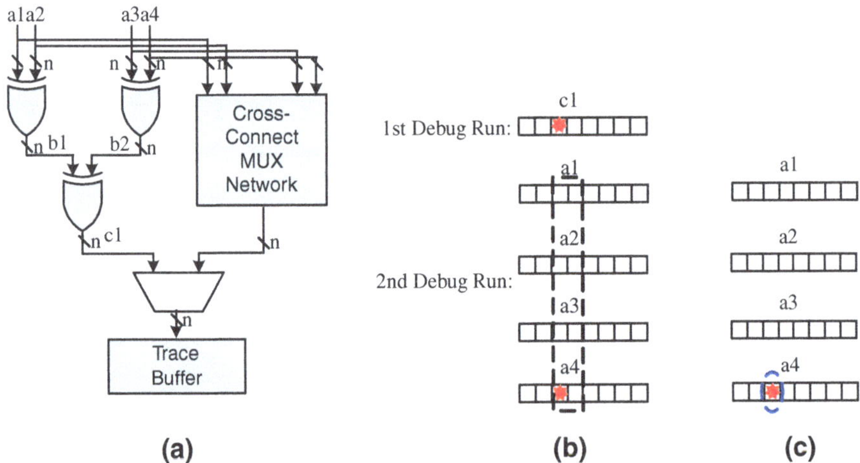

Fig. 8.4 A straightforward hybrid trace interconnection fabric. **a** Hybrid Network, **b** Trace Data, **c** Localized Error Evidence

[2] Running silicon for one second may take days for gate-level simulator to complete.

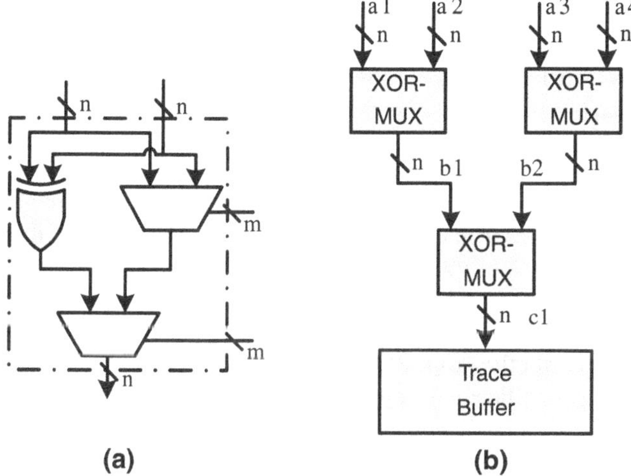

Fig. 8.5 Proposed trace interconnection fabric. **a** XOR-MUX Cell, **b** Proposed Fabric

For the same example shown earlier, after we localize the error evidence on the third-bit of trace data in the first debug run with the trace sigature, the fabric enables us to trace the original tapped signals directly and we can find the actual error evidence immediately in the second debug run (see Fig. 8.4b). While the number of debug runs can be reduced with this fabric design, it involves significant routing overhead as tapped signals that may be far from each other in the CUD need to be cross-connected. Moreover, if the "golden vectors" are not clean, we are not able to identify suspicious error signals via the XOR-based tracing in the first debug run and the efficiency of the debug solution is essentially the same as MUX-based fabric.

To tackle the above problem, we propose to have an XOR-MUX cell as the basic unit in our trace interconnection fabric. As depicted in Fig. 8.5a, from two groups of input signals, the cell is able to selectively transfer one group of signals or the parities from the two groups. Then, by replacing the MUX or XOR cell in Fig. 8.2 with the the newly-introduced XOR-MUX cell, we are able to obtain our proposed trace interconnection fabric with the same tree-like structure (see Fig. 8.5b) and the routing cost is similar to existing solutions.

There are two methods to configure the XOR-MUX cells: (i) control every single signal in the cell (i.e., $m = n$ in Fig. 8.5a), referred to as fine-grained control); (ii) control a group of signals with one selection signal ($m = 1$ in Fig. 8.5a, referred to as coarse-grained control). The former one involves more hardware overhead, but it enables more efficient silicon debug with higher flexibility; while the later one is the opposite. As the example shown in Fig. 8.5b, with coarse-grained control, we can only trace either $b1$ or $b2$ in each debug run, while with fine-grained control, we can select to trace some signals in $b1$ and some others in $b2$ concurrently. Note that, the configuration is again performed with JTAG interface and the routing overhead is acceptable. It also should be noted that how to select trace signals and how to group them in interconnection fabric is beyond the scope of this chapter. These problems can be resolved manually by designers. While in terms of automatic approach, signal selection can refer to [35, 42, 56, 62, 78], and signal grouping can refer to the structural analysis in [41].

To note, our proposed trace interconnection fabric design takes advantage of spatial compaction on trace data only (when XORs are used). On the other hand, it is also possible to compress trace data temporally using multiple input signature register (MISR) [4]. To be specific, the MISR-based trace compressor will periodically compress several cycles of trace data from interconnection fabric into a signature, and it can dramatically expand the tracing window. we can include MISR-based trace compressor into the proposed design to be able to compress trace data in the temporal domain. Together with the proposed flexible and X-tolerant interconnection fabric, we further develop an efficient flow with expanded tracing window for error evidence localization in trace-based debug.

8.3 Proposed Error Evidence Localization Methodology

With the proposed DfD design, we introduce our signal tracing methodology that enables efficient silicon debug in this section.

As discussed earlier, after a bug in the CUD is activated, it will take some time to propagate its error effects in the circuit and leave error evidences in one or more FFs. Suppose the trace time is sufficiently long and at least one of the tapped signals are left with such error evidence during tracing, the objective of our technique is to pinpoint the erroneous signal and the error occurrence cycle accurately. Moreover, as one debug run can cost significant runtime, it is beneficial to conduct the above process with as few debug runs as possible.

In the proposed error evidence localization methodology, we first configure the DfD design introduced in Sect. 8.2 to compress the trace data spatially and temporally. By doing this, we are with the largely extended capability to trace the internal behavior of the CUD. Consider the case that we have trace buffer with N_{period} (e.g., $64k$) depth and the MISR will periodically compress N_{depth} (e.g., $32k$) cycles of trace data into a signature, then we are able to monitor the CUD for $N_{period} \times N_{depth}$ (e.g., $2G$) cycles in total, which is capable to tackle those errors that take extremely long latency to manifest. By running the CUD once only, we can localize the error evidence existing on some tapped signals within the earliest monitored N_{period} (e.g., $32k$) cycles. Next we will configure the DfD design to bypass the trace compressor and utilize the proposed interconnection fabric flexibly to localize the exact error evidence within the earliest N_{period} (e.g., $32k$) cycles, detailed as follows.

First, let us consider the simple case that there exists clean "golden vectors". Under this circumstance, in the first debug run, we will configure the fabric to be an XOR network as shown in Fig. 8.6a, so that we can trace the erroneous parities from all tapped signals. Based on that, we will find some error evidences existing in tapped signals with the applied test vectors. To further localize it, we configure

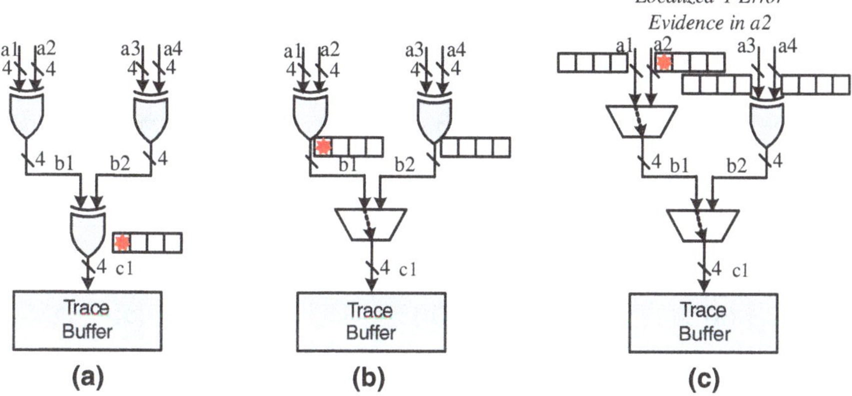

Fig. 8.6 Error evidence localization with clean "Golden Vectors". **a** 1st debug run, **b** 2nd debug run, **c** 3rd debug run

the XOR-MUX cell to selectively trace one group of parities at upper level (e.g., $b1$ as shown in Fig. 8.6b) in the next debug run. If the traced parities contains error, we will further trace the data at its upper level. As for this example, we notice $b1$ is with error and we will continue to configure the fabric to trace $a1$ as shown in Fig. 8.6c. This time, $a1$ is error-free and we can conclude that $a2$ is with error. The actual values can be calculated with $b1 \bigoplus a1$. By doing so, we will localize the exact one error evidence in $a2$ by taking d debug runs (d is the depth of tree-like fabric). If, however, we choose to trace $b2$ in the second debug run, we will find that parities in $b2$ are correct, which means all signals at the upper levels of $b2$ are also correct. Consequently, we will still trace the upper level signals of $b1$ (i.e., $a1$ or $a2$) and we will localize the same error evidence in $a2$ in the third debug run. Therefore, no matter that we selectively trace the signal group with erroneous parity or not, the above binary search-like procedure guarantees to localize error evidence with d debug runs. To note, the actual structure may not be the ideal case that each group are with the same amount of tapped signals and each sub-tree are with the same depth. However, the proposed signal tracing methodology is able to work in the same manner.

Next, let us consider the case when X-bits exist in "golden vectors". Unlike XOR-based fabric, our proposed fabric is able to avoid signature corruption naturally by blocking X-bits during their transfer. For the case with fine-grained control fabric that can direct every signal, as the example shown in Fig. 8.7b, $a1$ and $a2$ are with X-bits in "golden vectors", then we will configure the XOR-MUX cell to transfer signals with known values only during signal tracing, so that X-bits would have no impact. For this particular example, we are able to tolerate all the X-bits before they corrupt any possible error evidence. Suppose that more signals are with X-bits, we may not be able to block them at the first level (e.g., the same bit position in $a1$ and $a2$ are with X-bits), but eventually we can mask them at lower levels (e.g., in $b1$). For

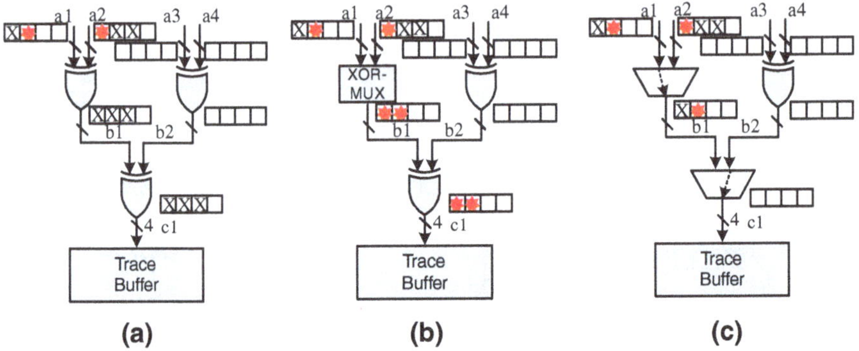

Fig. 8.7 Error evidence localization with X-bits in "Golden Vectors". **a** XOR-Based Fabric, **b** Proposed Fabric Fine-Grained Control, **c** Proposed Fabric with Coarse-Grained Control

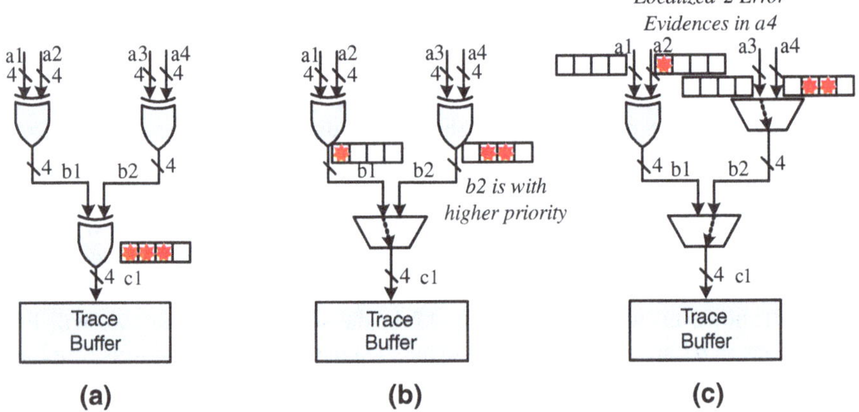

Fig. 8.8 Collecting more error evidences during signal tracing. **a** XOR-Based Fabric, **b** Proposed Fabric Fine-Grained Control, **c** Proposed Fabric with Coarse-Grained Control

the case with coarse-grained control fabric that can only direct every group of signals together, we can also block the X-bits in one group by configuring the XOR-MUX cells to transfer the other group, as shown in Fig. 8.7c. However, the error evidences in the masked group will not be observable and affect the error detection quality.

If the bug manifests itself on multiple tapped signals, it is beneficial to collect more error evidences during signal tracing, as we only need to study the possible root-cause region that affects all error evidences. For the simple case that we can use fine-grained control to direct every signal, we can directly adopt the earlier approach so that all error evidences are finally collected. This is achieved by repeatedly using the error localization flow for each bit of trace data. As the example shown in Fig. 8.8, the proposed approach is able to collect three error evidences in $a2$ and $a4$ because these evidences are on different bit position in trace data.

If we are equipped with trace interconnection fabric with coarse-grained control only, to collect as many error evidences as possible, the overall flow of procedure is similar to the one discussed earlier (see Fig. 8.6), and the main difference is how to determine the signals to trace in the next debug run. In previous method, we simply decide to trace one signal group on the upper level if the current signal group is with any erroneous parity. Now, for the case that both signal group are with erroneous parity, we will give higher priority to trace the upper level of the signal group with more erroneous parities, so that eventually more error evidences are likely to be collected. As the example shown in Fig. 8.8b, by tracing $b1$ in the second debug run and calculating $b2$ from $b1 \bigoplus a1$, we find $b1$ and $b2$ is with one and two erroneous parities, respectively. We will then choose to trace the upper level of $b2$ as it is with 1 more erroneous parity than $b1$. Finally, we can localize on $a4$ with two error evidences.

It should be noted that our approach only takes advantage of spatial compaction on trace data. Previously, [4] proposed to temporally compact trace data by using multiple input signature register. As two orthogonal solutions, these can be integrated together so that the observation on trace-based silicon debug is dramatically increased. One important application of the solution is to resolve the repeatable electrical error taking millions of cycles to expose, which remains to be the most challenging problem in silicon debug [43, 53].

8.4 Experimental Results

8.4.1 Experimental Setup

We conduct experiments on several large ISCAS'89 and IWLS'05 benchmark circuits to evaluate the effectiveness of the proposed solution against existing ones. In terms of the interconnection fabrics, we consider MUX-based fabric, XOR-based fabric and the proposed one with coarse-grained control and fine-grained control, respectively. For circuits s38417, s38584 and *usb*, we randomly select 300 tapped signals, while for larger circuits *des* and *ethernet* we tap 1000 signals.

The corresponding signal tracing solutions work as follows. For the one equipped with MUX-based fabric, it simply traces different subsets of signals in each debug run until any error evidence is detected. For the one using XOR-based fabric, it uses one single debug run to trace compacted signatures from all tapped signals. While for the proposed trace interconnection fabric, we try to collect as many error evidences as possible with coarse-grained control and fine-grained control, respectively.

To evaluate these tracing methods' capability on error detection, we randomly generate 1000 errors for each circuit, and each time, we inject one of the errors into the original netlist to obtain the erroneous netlist. Simulation with $16k$ cycles is then conducted to dump the actual states. As no dedicated trigger mechanism is used in our experiments, the state dumping starts from the beginning of the simulation.

Table 8.1 Experimental results on error detection quality evaluation (Buffer Width = 8)

Circuit	# of acti. error	# of det. error	# of debug runs				# of acual/sus. error evidences			
			XOR	MUX	Pro.C	Pro.F	XOR	MUX	Pro.C	Pro.F
s38417	751	447	1	27.6	3.23	3.64	2.58/73.1	1.03/1.03	1.12/1.12	2.24/2.24
s38584	825	684	1	19.1	4.47	5.06	3.41/91.3	1.05/1.05	1.12/1.12	2.94/2.94
usb	634	222	1	33.5	2.20	2.32	1.78/57.4	1.00/1.00	1.05/1.05	1.70/1.70
des	879	854	1	22.8	6.95	6.98	32.2/489.4	1.15/1.15	1.72/1.72	7.28/7.28
ethernet	227	88	1	120.3	1.61	1.61	1.89/167.6	1.00/1.00	1.01/1.01	1.43/1.43

Table 8.2 Experimental results on error detection quality evaluation (Buffer Width = 16)

Circuit	# of acti. error	# of det. error	# of debug runs				# of acual/sus. error evidences			
			XOR	MUX	Pro.C	Pro.F	XOR	MUX	Pro.C	Pro.F
s38417	751	447	1	14.2	2.85	3.23	2.77/45.5	1.08/1.08	1.23/1.23	2.65/2.65
s38584	825	684	1	9.9	3.89	4.39	3.57/59.2	1.20/1.20	1.12/1.12	3.70/3.70
usb	634	222	1	17.2	1.99	2.11	1.86/33.3	1.05/1.05	1.10/1.10	1.91/1.91
des	879	854	1	11.7	6.12	6.12	32.1/482.3	1.56/1.56	2.68/2.68	13.2/13.2
ethernet	227	88	1	60.6	1.53	1.53	2.06/95.2	1.04/1.04	1.04/1.04	2.06/2.06

Table 8.3 Experimental results on error detection quality evaluation (Buffer Width = 32)

Circuit	# of acti. error	# of det. error	# of debug runs				# of acual/sus. error evidences			
			XOR	MUX	Pro.C	Pro.F	XOR	MUX	Pro.C	Pro.F
s38417	751	447	1	7.76	2.50	2.79	2.96/25.9	1.19/1.19	1.41/1.41	3.02/3.02
s38584	825	684	1	5.50	3.30	3.73	4.33/38.9	1.40/1.40	1.80/1.80	4.56/4.56
usb	634	222	1	9.40	1.77	1.89	2.02/15.6	1.12/1.12	1.25/1.25	2.09/2.09
des	879	854	1	6.38	5.27	5.27	32.6/442.1	1.96/1.96	4.20/4.20	22.1/22.1
ethernet	227	88	1	31.3	1.44	1.44	2.02/53.2	1.07/1.07	1.07/1.07	1.85/1.85

These states are compared against the "golden vector" obtained from the simulation with the original netlist to get the propagated error evidences, by finding the difference between actual states and "golden vector". Finally, with different signal tracing methods, a bug is regarded to be detected if the signal tracing solution finds any error evidence.

8.4.2 Results and Discussion

Tables 8.1, 8.2 and 8.3 present the experimental results on error detection quality of various tracing solutions when trace buffer width is 8, 16 and 32, respectively. In these tables, Column 2 and Column 3 shows the number of activated errors and the number of detected errors. We can observe that, in average, 66.3 % of the injected errors are activated in the experiment, and the conditions to activate other injected errors are not met with simulation for $16k$ cycles. Within the activated errors, around 70 % are detected with trace-based debug. This is due to the fact that we only tap a small portion of the signals in the circuits (9.5–20.5 %) and hence we will miss the error evidences if the bug propagates to other signals. This is also because the tapped signals in our experiment are randomly selected and the detection quality cannot be guaranteed.[3]

Columns 4–7 present the average number of debug runs (denoted by "# of Debug Runs") for different tracing solutions. Among them, MUX-based fabric (denoted by "MUX" in Column 5) requires the largest number of debug runs (26.5 on average), which means the required tracing time is more than other solutions. With the increase of buffer width, however, the associated tracing capability is enhanced and the required debug runs for this solution is decreased in linear manner. XOR-based fabric (denoted by "XOR" in Column 4) conducts signal tracing for one debug run for all cases. While for both of our proposed solutions with coarse-grained control (denoted by "Pro.C" in Column 6) and fine-grained control (denoted by "Pro.F" in Column 7), only 3.27 and 3.47 debug runs are needed for detecting error evidences. Both are much less than MUX-based fabrics. This is because for the case that no error evidence exists, the proposed solution will stop at the first debug run as no erroneous parity is found, while the one with MUX-based fabric takes many debug runs for tracing every tapped signal to make the conclusion. On the other case that some error evidences do exist, the number of required debug runs is also small as it is strictly bounded by the depth of tree-like interconnection fabric.

The average number of actual/supicous error evidences obtained from various tracing solutions is shown in Column 8–12 (denoted by "# of Acual/Sus. Error Evidence"). Clearly, the tracing solution with MUX-based fabric and our proposed solutions are able to guarantee every suspicious error evidence is the actual one. While for the method with XOR network, only 5.9 % of the suspicious error evidences are the actual ones, which results in higher diagnosis effort. In terms of the capability of collecting actual error evidence, the proposed solution with coarse-grained control and the one with fined-grained control collect 29.1 and 303 % more error evidences than MUX-based fabric, respectively.

The next experiment evaluates the error detection quality of various tracing solutions when X-bits exist in provided "golden vectors". Here, we randomly inject a certain ratio of X-bits in trace data to compare the X-tolerant capability with different tracing solutions for benchmark circuit s38417. As shown in Fig. 8.9, by applying different tracing solutions on the original trace data without X-bits, the number of detected error are similar. After that, as more X-bits are injected, the error detection quality with the MUX-based fabric is not significantly reduced, because the method uses many debug runs to trace the data from different signals, and any remaining error evidences (i.e., not X-bits) will still be detected. However, for the tracing solution using XOR network, the X-bits will be easily corrupted and its error detection

[3] From this perspective, it is essential to develop high-quality trace signal selection methods to increase error detection capability, but this is out of the scope of this work.

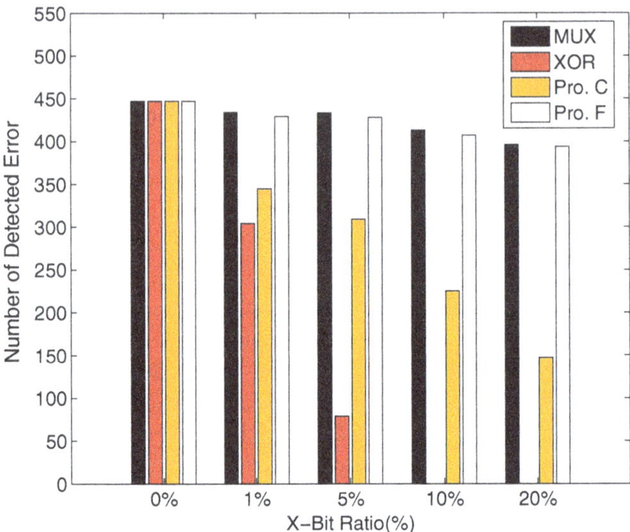

Fig. 8.9 Error detection quality evaluation with X-Bits on s38417

Table 8.4 Area overhead of different tracing solutions

Buffer width	Δ_{XOR} (%)	$\Delta_{Pro.C.}$ (%)	$\Delta_{Pro.F.}$ (%)
8	−0.09	3.88	9.90
16	0.07	1.86	5.13
32	0.06	0.91	2.63

$\Delta = \frac{Extra\ Hardware\ Cost}{MUX\text{-}based\ Hardware\ Cost} \times 100\,\%$

quality is greatly reduced with the growth of X-bits. As shown in Fig. 8.9, the method will not detect any error if the ratio of X-bits is greater than 10 %. On the other hand, our proposed solution shows its strong capability on tolerating X-bits. With our fine-grained control tracing method, the error detection quality is slightly affected with X-bits. Even for the case with 20 % X-bits in trace data, the number of detected errors is almost as high as the one with MUX-based fabric with much fewer debug runs. Even for our coarse-grained control tracing method that consumes less DfD overhead, its error detection quality is slowly dropped by the growth of X-bits, and remains to be considerably high with 5 % X-bits in trace data.

Finally, to evaluate the area cost for the different tracing solutions, we implement and synthesize the DfD hardware for circuit *des* using a commercial tool for the conventional MUX-based fabric, XOR-based fabric and the proposed one with coarse-grained and fine-grained control, respectively. The number of tapped signal is fixed as 1000. As shown in Table 8.4, we compare the hardware cost of the other solutions against the conventional MUX-based fabric, assuming the trace buffer depth to be 16k and the trace buffer width to be 8, 16 and 32, respectively. As indicated in Column 2, XOR-based fabric consumes similar hardware cost as MUX-based fabric. Meanwhile, the proposed solution with coarse-grained control (in Column 3) takes less than 4 % extra cost, which is quite small. Even for the solution equipped with fine-grained control (in Column 4), the extra overhead is kept to be less than 10 %. This is due to the fact that interconnection fabric only takes a small portion of the DfD area in the trace-based debug infrastructure, and although the proposed solutions require more hardware cost on interconnection fabric, the overall DfD area will not increase significantly. On the other hand, as the trace buffer width increases, the tree-like interconnection fabric of proposed solution requires less XOR-MUX cells to conduct the signal concentration and hence the corresponding hardware cost further decreases.

8.5 Conclusion

In this chapter, we propose a flexible low-cost trace interconnection fabric design and the corresponding signal tracing method that is able to localize the error with high accuracy within a few debug runs. In addition, the solution is with strong X-tolerant capability. Experimental results on benchmark circuits demonstrate the efficacy of proposed technique.

Chapter 9
Conclusion

The ever-increasing complexity of modern circuits challenges our ability to guarantee their correctness. Consequently, various errors escape the pre-silicon verification process and manifest themselves after the tape-out. To resolve the problem, several post-silicon validation techniques are proposed by both academia and industry for eliminating design bugs before IC products are shipped to customers. In order to reduce time-to-market and to avoid re-spin cost, DfD techniques have been widely accepted for enhancing the debug capability in post-silicon validation. In this book, we introduce several novel techniques to exhaustively explore this problem.

We develop a novel signal selection technique in Chap. 3 to maximize the visibility of design errors during signal tracing. The solution is further improved for working with "golden vector" from high-level simulation and assertion-based debug. In Chap. 4, to improve the ability for detecting these errors, we further introduce a multiplexed signal tracing strategy, in which we divide the tracing procedure in each debug run into a few periods and trace different sets of signals in each period. The technique is supported with a signal grouping algorithm and DfD hardware. On the other hand, to detect speedpath-related electrical errors, in Chap. 5 we propose a signal selection solution together with a trace qualification technique. We also introduce several low-cost interconnection fabrics to transfer trace data while improving the debug effectiveness in post-silicon validation. The first work introduced in Chap. 6 is to reuse the existing test channel for real-time trace data transfer so that DfD cost is reduced. We then develop a novel interconnection fabric design as well as the associated optimization technique in Chap. 7 to achieve high debug flexibility with minimized hardware cost. Moreover, in Chap. 8 we develop an unknown bits tolerant interconnection fabric, with which we are able to develop a systematic signal tracing procedure to automatically localize erroneous signals with a few debug runs. Experimental results demonstrate the effectiveness of the proposed techniques.

There are several interesting and important research topics for future work. The quality of trace signal selection introduced in Chap. 3 highly depends on the metric. However, the accuracy of the proposed metric cannot be very high because we do not have an efficient solution to accurately obtain the metric. As the result, studying

X. Liu and Q. Xu, *Trace-Based Post-Silicon Validation for VLSI Circuits*,
Lecture Notes in Electrical Engineering, DOI: 10.1007/978-3-319-00533-1_9,

above problem will be useful to improve the effectiveness on the trace signal selection technique. Chapter 4 introduces the idea of multiplexed tracing. To further explore it, we can focus on two problems: (i) How to select trace signals during design phase to facilitate better design error detection? (ii) Based on the error detection information that only provides *part-view* of the CUD, how to conduct effective diagnosis to root-cause the error. Chapter 6 describes the method that reuses test channel for real-time trace data transfer. As currently the SoC test architecture design and optimization focuses on test time reduction, the resultant test architecture thus may not be efficient in terms of debug data transfer. Then the next relevant problem is how to take the cores' debug requirements into consideration during test architecture design so that the TAM bandwidth utilization rate for debug can be enhanced. Besides above future topics relevant to the book, we are facing many open problems in post-silicon validation. One key problem is that we need generic evaluation metrics to measure the quality of debug solutions (e.g., similar to "fault coverage" in manufacturing test or "state coverage" in simulation). In addition, most existing works in this area focused on debugging logic errors that are easy to be repeated. A more challenging problem is to debug those electrical errors that takes long time to manifest themselves in nondeterministic manner.

In summary, post-silicon validation becomes a vital step in the design flow of VLSI circuits. Although many ad-hoc solutions have been proposed in practice, the validation difficulty from escalating complexity and inaccurate model motivates the need for automatic and algorithmic solutions. In this book, we investigate several key research problems in this area. The proposed algorithmic solutions together with new introduced DfD hardware provide an important contribution on the automatic techniques in post-silicon validation.

References

1. Abramovici, M. (2008). In-system silicon validation and debug. *IEEE Design & Test of Computers*, *25*(3), 216–223.
2. Abramovici, M., Bradley, P., Dwarakanath, K., Levin, P., Memmi, G., & Miller D. (2006). A reconfigurable design-for-debug infrastructure for SoCs. *Proceedings ACM/IEEE Design Automation Conference (DAC)* (pp. 7–12), July 2006.
3. Altera Inc. Design debugging using the signalTap II embedded logic analyzer. http://www.altera.com
4. Anis, E., & Nicolici, N. (2007). Low cost debug architecture using lossy compression for silicon debug. *Proceedings Design, Automation, and Test in Europe (DATE)* (pp. 1–6).
5. Anis, E., & Nicolici, N. (2007). On using lossless compression of debug data in embedded logic analysis. *Proceedings IEEE International Test Conference (ITC)* (pp. 1–10).
6. ARM Ltd. Embedded trace macrocell architecture specification. http://www.arm.com
7. ARM Ltd. How CoreSight technology gets higher performance, more reliable product to market quicker. http://www.arm.com
8. Bastani, P., Killpack, K., Wang, L. C., & Chiprout, E. (2008). Speedpath prediction based on learning form a small set of examples. *Proceedings ACM/IEEE Design Automation Conference (DAC)* (pp. 217–222).
9. Basu, K., & Mishra, P. (2011). Efficient trace signal selection for post silicon validation and debug. *Proceedings International Conference on VLSI Design.*
10. Boule, M., Chenard, J., & Zilic, Z. (2007) Assertion checkers in verification, silicon debug and in-field diagnosis. *Proceedings International Symposium on Quality of Electronic Design (ISQED)* (pp. 613–620).
11. Boule, M., Chenard, J., & Zilic, Z. (2007). Debug enhancements in assertion-checker generation. *IEEE Design & Test of Computers*, *1*(6), 669–677.
12. Brglez, F. (1984). On testability analysis of combinational networks. *Proceedings IEEE Symposium on Circuits and Systems* (pp. 221–225).
13. Callegari, N., Wang, L. C., & Bastani, P. (2009). Speedpath analysis based on hypothesis pruning and ranking. *Proceedings ACM/IEEE Design Automation Conference (DAC)* (pp. 346–351).
14. Chang, K. H., Bertacco, V., & Markov, I. L. (2005). Simulation-based bug trace minimization with BMC-based refinement. *Proceedings International Conference on Computer-Aided Design (ICCAD)* (pp. 1045–1051).
15. Chang, K. H., Markov, I., & Bertacco, V. (2007). Fixing design errors with counterexamples and resynthesis. *Proceedings IEEE Asia South Pacific Design Automation Conference (ASP-DAC)* (pp. 944–949).
16. Clos, C. (1953). A study of non-blocking switching networks. *Bell System Technical Journal*, *32*, 406–424.

X. Liu and Q. Xu, *Trace-Based Post-Silicon Validation for VLSI Circuits*, Lecture Notes in Electrical Engineering, DOI: 10.1007/978-3-319-00533-1,

17. Cloutier, R. J., Goossens, K., Basten, T., Radulescu, A., & Boon, A. (2006). Transaction monitoring in networks on chip: The on-chip run-time perspective. *Proceedings of Symposium on Industrial Embedded Systems* (pp. 1–10).
18. de Micheli, G. (1994). *Synthesis and optimization of digital circuits*. New York: McGraw-Hill International Editions.
19. Doering, R., & Nishi, Y. (2007). *Handbook of semiconductor manufacturing technology*. Upper Saddle River: Prentice-Hall.
20. Goel, S. K., Chiu, K., Marinissen, E. J., Nguyen, T., & Oostdijk, S. (2004). Test infrastructure design for the nexperia™ home platform PNX8550 system chip. *Proceedings Design, Automation, and Test in Europe (DATE)* (pp. 108–113).
21. Gramm, J., Guo, J., Huffner, F., & Niedermeier, R. (2006). Data reduction, exact, and heuristic algorithms for clique cover. *Workshop on Algorithm Engineering and Experiments* (pp. 8–94).
22. Grotker, T., Liao, S., Martin, G., & Swan, S. (2002). *System design with systemC*. New York: Springer-Verlag.
23. Hopkins, A. B., & McDonald-Maier, K. D. (2006). Debug support for complex systems on-chip: A review. *In IEE Proceedings of Computers and Digital Techniques*(pp. 197–207).
24. Hopkins, A. B. T., & McDonald-Maier, K. D. (2006). Debug support for complex systems on-chip: A review. *IEE Proceedings, Computers and Digital Techniques, 153*(4), 197–207.
25. Hopkins, A. B. T.,& McDonald-Maier, K. D. (2007). Trace algorithms for deeply integrated complex and hybrid SoCs. *NASA/ESA Conference on Adaptive Hardware and Systems* (pp. 641–646).
26. Hsu, Y. C., Tsai, F., Jong, W., & Chang, Y. T. (2006). Visibility enhancement for silicon debug. *Proceedings ACM/IEEE Design Automation Conference (DAC)* (pp 13–18), July 2006.
27. Huang, Y., & Cheng, W. (2003). Using embedded infrastructure IP for SOC post-silicon verification. *Proceedings ACM/IEEE Design Automation Conference (DAC)* (pp. 674–677).
28. Hwang, F. (1999). *The mathematical theory of nonblocking switching networks*. River Edge: World Scientific Publishing Company.
29. Jain, S. K., & Agrawal, V. D. (1984). Statistical fault analysis. *IEEE Design & Test, 2*(1), 38–44.
30. Josephson, D. D. (2002). The manic depression of microprocessor debug. *Proceedings IEEE International Test Conference (ITC)* (pp. 657–663), October 2002.
31. Josephson, D. D., & Gottlieb, B. (2001). Debug methodology for the McKinley processor. *Proceedings IEEE International Test Conference (ITC)* (pp. 451–460), October 2001.
32. Josephson, D. D., & Gottlieb, B. (2004). The crazy mixed up world of silicon debug. *Proceedings IEEE Custom Integrated Circuits Conference (CICC)* (pp. 665–670), October 2004.
33. Kern, C., & Greenstreet, M. R. (1999). Formal verification in hardware design: A survey. *ACM Transactions on Design Automation of Electronic Systems, 4*(2), 123–193.
34. Ko, H. F., Kinsman, A. B., & Nicolici, N. (2008). Distributed embedded logic analysis for post-silicon validation of SOCs. *Proceedings IEEE International Test Conference (ITC)* (pp 1–10).
35. Ko, H. F., & Nicolici, N. (2008). Automated trace signals identification and state restoration for improving observability in post-silicon validation. *Proceedings Design, Automation, and Test in Europe (DATE)* (pp. 1298–1303).
36. Ko, H. F., & Nicolici, N. (2009). Algorithms for state restoration and trace-signal selection for data acquisition in silicon debug. *IEEE Transactions on Computer-Aided Design, 28*(2), 285–297.
37. Kropf, T. (2000). *Introduction to formal hardware verification*. New York: Springer-Verlag.
38. Lai, C., Yang, F., Kao, C., & Huang, I. (2009). A trace-capable instruction cache for cost efficient real-time program tracc compression in SoC. *Proceedings ACM/IEEE Design Automation Conference (DAC)* (pp. 136–141).
39. Leatherman, R., & Stollon, N. (2005). An embedded debugging architecture for SOCs. *IEEE Potentials, 24*(1), 12–16.
40. Liu, X., & Xu, Q. (2008). On reusing test access mechanisms for debug data transfer in SoC post-silicon validation. *Proceedings IEEE Asian Test Symposium (ATS)* (pp. 303–308).

41. Liu, X., & Xu, Q. (2009). Interconnection fabric design for tracing signals in post-silicon validation. *Proceedings ACM/IEEE Design Automation Conference (DAC)* (pp. 352–357).
42. Liu, X., & Xu, Q. (2009). Trace signal selection for visibility enhancement in post-silicon validation. *Proceedings Design, Automation, and Test in Europe (DATE)* (pp. 1338–1343).
43. Liu, X., & Xu, Q. (2010). On signal tracing for debugging speedpath-related electrical errors in post-silicon validation. *Proceedings IEEE Asian Test Symposium (ATS)* (pp. 243–248).
44. Liu, X., & Xu, Q. (2011). On multiplexed signal tracing for post-silicon debug. *Proceedings Design, Automation, and Test in Europe (DATE)*.
45. Marinissen, E. J., Arendsen, R., Bos, G., Dingemanse, H., Lousberg, M., & Wouters, C. (1998). A structured and scalable mechanism for test access to embedded reusable cores. *Proceedings IEEE International Test Conference (ITC)* (pp. 284–293), October 1998.
46. MIPS Technologies Inc. EJTAG trace control block specification. http://www.mips.com
47. Nakamura, S., & Masson, G. M. (1982). Lower bounds on crosspoints in concentrators. *IEEE Transactions on Computers, C-31*(12), 1173–1179.
48. Narasimha, M. J. (1994). A recursive concentrator structure with applications to self-routing switching networks. *IEEE Transactions on Communications, 42*(234), 896–898.
49. Nataraj, N., Lundquist, T., & Shah, K. (2003). Fault localization using time resolved photon emission and STIL waveforms. *Proceedings IEEE International Test Conference (ITC)* (pp. 254–263).
50. Navabi, Z. (1997). *VHDL: Analysis and modeling of digital systems*. New York: McGraw-Hill Inc.
51. Z. Navabi. Verilog Digital System Design (Professional Engineering). McGraw-Hill Inc., 1999.
52. Paniccia, M., Eiles, T., Rao, V. R. M., & Yee, W. M. (1998). Novel optical probing technique for flip chip packaged microprocessors. *Proceedings IEEE International Test Conference (ITC)* (pp. 740–747).
53. Park, S. B., & Mitra, S. (2008). IFRA: Instruction footprint recording and analysis for post-silicon bug localization in processors. *Proceedings ACM/IEEE Design Automation Conference (DAC)*.
54. Pippenger, N. (1974). On the complexity of strictly nonblocking concentration networks. *IEEE Transactions on Communications, 22*(11), 1890–1892.
55. Prabhakar, S., & Hsiao, M. S. (2009). Using non-trivial logic implications for trace buffer-based silicon debug. *Proceedings IEEE Asian Test Symposium (ATS)* (pp. 131–136).
56. Prabhakar, S., & Hsiao, M. S. (2010). Multiplexed trace signal selection using non-trivial implication-based correlation. *Proceedings International Symposium on Quality of Electronic Design (ISQED)* (pp. 697–704).
57. Quinton, B. R., & Wilton, S. J. E. (2005). Concentrator access networks for programmable logic cores on SoCs. *Proceedings International Symposium on Circuits and Systems (ISCAS)* (pp. 45–48).
58. Rashinkar, P., Paterson, P., & Singh, L. (2002). *System-on-a-chip verification: Methodology and techniques*. New York: Kluwer Academic Publishers.
59. Rootselaar, G., & Vermeulen, B. (1999). Silicon debug: Scan chains alone are not enough. *Proceedings IEEE International Test Conference (ITC)* (pp. 892–902), September 1999.
60. Schlangen, R., Kerst, U.,& Boit, C. (2007). New circuit edit and probing options directly to FET device on ultra thin silicon backside processed by focused ion beam. *Proceedings IEEE International Workshop on Silicon Debug and Diagnosis* (pp. 328–333).
61. Semiconductor Industry Association (SIA). (2003). The international technology roadmap for semiconductors (ITRS): 2003 edition. http://public.itrs.net/Files/2003ITRS/Home2003.htm
62. Shojaei, H., & Davoodi, A. (2010). Trace signal selection to enhance timing and logic visibility in post-silicon validation. *Proceedings International Conference on Computer-Aided Design (ICCAD)* (pp. 168–172).
63. Stollon, N., Leatherman, R., Ableidinger, B., Edgar, E. Multi-core embedded debug for structured ASIC systems. http://www.fs2.com/
64. Stollon, N., Leatherman, R., Ableidinger, B., & Edgar, E. (2004). Multi-core embedded debug for structured ASIC Systems. *Proceedings of DesignCon*.

65. Tang, S., & Xu, Q. (2008). In-band cross-trigger event transmission for transaction-based debug. *Proceedings Design, Automation, and Test in Europe (DATE).*
66. Vallett, D. (1997). IC failure analysis: The importance of test and diagnostics. *IEEE Design & Test of Computers, 14*(3), 76–82.
67. Varma, P., & Bhatia, S. (1998). A structured test re-use methodology for core-based system chips. *Proceedings IEEE International Test Conference (ITC)* (pp. 294–302), Washington, DC, Oct. 1998.
68. Vermeulen, B., & Goel, S. K. (2002). Design for debug: Catching design errors in digital chips. *IEEE Design & Test of Computers, 19*(3), 37–45.
69. Vermeulen, B., Oostdijk, S., & Bouwman, F. (2001). Test and debug strategy of the PNX8525 Nexperia^{TM} digital video platform system chip. *Proceedings IEEE International Test Conference (ITC)* (pp. 121–130), Baltimore, MD, Oct. 2001.
70. Vermeulen, B., Waayers, T., & Bakker, S. (2002). IEEE 1149.1-compliant access architecture for multiple core debug on digital system chips. *Proceedings IEEE International Test Conference (ITC)* (pp. 55–63), Baltimore, MD, Oct. 2002.
71. Vermeulen, B., Waayers, T., & Goel, S. K. (2002). Core-based scan architecture for silicon debug. *Proceedings IEEE International Test Conference (ITC)* (pp. 638–647), October 2002.
72. Vermeulen, B., Waayers, T., & Goel, S. K. (2007). Transaction-based communication-centric debug. *ACM/IEEE International Symposium on Networks-on-Chip (NOCS)*, May 2007.
73. Vishnoi, A., Panda, P., & Balakrishnan, M. (2009) Cache aware compression for processor debug suppport. *Proceedings Design, Automation, and Test in Europe (DATE).*
74. Xilinx Inc. Chipscope pro software and cores user guide. http://www.xilinx.com
75. Xu, Q. & Liu, X. (2010). On signal tracing in post-silicon validation. *Proceedings IEEE Asia South Pacific Design Automation Conference (ASP-DAC)* (pp. 262–267).
76. Xu, Q., & Nicolici, N. (2005). Resource-constrained system-on-a-chip test: A survey. *IEE Proceedings, Computers and Digital Techniques, 152*(1), 67–81.
77. Yang, J.-S., & Touba, N. A. (2008). Enhancing silicon debug via periodic monitoring. *Proceedings IEEE International Symposium on Defect and Fault Tolerance in VLSI Systems (DFT)* (pp. 125–133).
78. Yang, J.-S., & Touba, N. A. (2009). Automated selection of signals to observe for efficient silicon debug. *Proceedings IEEE VLSI Test Symposium (VTS)* (pp. 79–84).
79. Zorian, Y., Marinissen, E. J., & Dey, S. (1999). Testing embedded-core-based system chips. *IEEE Computer, 32*(6), 52–60.

Index

X. Liu and Q. Xu, *Trace-Based Post-Silicon Validation for VLSI Circuits*,
Lecture Notes in Electrical Engineering, DOI: 10.1007/978-3-319-00533-1,

Zeitfracht Medien GmbH
Ferdinand-Jühlke-Straße 7
99095 Erfurt, Deutschland
produktsicherheit@kolibri360.de